A Contextual Exploration of Phytomedicines' Development in Africa

A Contextual Exploration of Phytomedicines' Development in Africa

Obi Peter Adigwe and
Kofi Busia

Library of Congress Control Number:		2022922903
ISBN:	Hardcover	978-1-6641-1851-5
	Softcover	978-1-6641-1852-2
	eBook	978-1-6641-1853-9

Print information available on the last page.

Rev. date: 01/16/2023

To order additional copies of this book, contact:
Xlibris
UK TFN: 0800 0148620 (Toll Free inside the UK)
UK Local: (02) 0369 56328 (+44 20 3695 6328 from outside the UK)
www.Xlibrispublishing.co.uk
Orders@Xlibrispublishing.co.uk
848934

Edited by

Obi Peter Adigwe
CEO/Director General, National Institute for Pharmaceutical Research and Development (NIPRD), Federal Ministry of Health, Garki, Abuja FCT, Nigeria

Kofi Busia
Editor-in-Chief of the Journal of Herbal Medicine, Germany.
Immediate Past Director of Healthcare Services, West African Health Organisation, Burkina Faso.
Principal Postgraduate Supervisor, Faculty of Medicine, Lincoln University College, Malaysia.

CONTENTS

CONTRIBUTORS

Irene Aasam Aabeinir is a medical herbalist intern at the Centre for Plant Medicine Research. She is an enthusiast and advocate for quality healthcare, and her interest is in clinical research.

Emmanuel Adase is a technologist at the Department of Production at the Centre for Plant Medicine Research. He specialises in quality control processes and acts as a research assistant in the formulation and evaluation of phytomedicines.

Obi Peter Adigwe is the CEO/Director General of the National Institute for Pharmaceutical Research and Development (NIPRD), Federal Ministry of Health, Garki, Abuja FCT, Nigeria.

Isaac Kingsley Amponsah is an associate professor of pharmacognosy at the Kwame Nkrumah University of Science and Technology, Kumasi, Ghana. He is an expert in the physicochemical analysis and spectroscopic characterization of natural products as well as in monograph development. He studied good agricultural and collection practices for medicinal plants-WHO model from the Hanoi School of Pharmacy, Vietnam.

Mary-Ann Archer is a pharmacist with expertise in pharmaceutical quality systems: pharmaceutical technology, drug delivery systems and phytopharmaceuticals. She focuses her research activities on developing suitable dosage forms and novel drug delivery systems for phytopharmaceuticals as well as the innovative use of natural products as pharmaceutical excipients.

Mavis Boakye-Yiadom is a chief medical herbalist (clinician), research officer and head of the Clinical Research Department of the Centre for Plant Medicine Research. She specialises in clinical, phytochemistry, pharmacology and phytomedicine evaluation research.

Kofi Busia is the immediate past director of healthcare services at the West African Health Organisation, Burkina Faso; the editor-in-chief of the Journal of Herbal Medicine, UK; and a principal postgraduate supervisor of the Faculty of Medicine, Lincoln University College, Malaysia.

Peter Bai James is a postdoctoral research fellow at the National Centre for Naturopathic Medicine at Southern Cross University, Australia, and an adjunct lecturer at the Faculty of Pharmaceutical Sciences, College of Medicine and Allied Health Sciences, University of Sierra Leone. He is a member of the advisory committee on the traditional medicine program at the West African Health Organisation.

Julius Ossy Muganga Kasilo is the regional adviser for traditional medicine and team leader for health technologies at the WHO/AFRO, the medicines regulatory officer at WHO, Geneva; and the focal point for the Joint UNIDO-WHO Project on 'Enhancing the provision of personal protective equipment in Africa: Strengthening the resilience of the industrial and healthcare sectors to COVID-19 and future pandemics'.

Charles Katy is a consultant in human sciences, traditional medicine and indigenous knowledge as well as a facilitator of community cultural networks.

Doris Kumadoh is a research officer in the Department of Pharmaceutics and Quality Control and the head of the production department of the Centre for Plant Medicine Research. She specialises in phytomedicine formulations, evaluation, standardization, stability studies and pharmaceutical technology.

Michael Odoi-Kyene is a research scientist at the Department of Pharmaceutics and Quality Control of the Centre for Plant Medicine Research. His expertise includes analytical chemistry, herbal products formulation, quality control analysis, and nano drug delivery methods.

Nanlop Adenike Ogbureke is a monitoring and evaluation expert and senior specialist advisor to the director general of the West African Health Organisation, Bobo Dioulasso, Burkina Faso.

Genevieve Naana Yeboah is a research officer at the Centre for Plant Medicine Research, with research interests in the development and characterisation of oral and rectal dosage forms from medicinal plants and exploration of naturally occurring excipients as drug delivery vehicles and topical drug delivery.

Zhou Jephias Redemptor is a Zimbabwean who studied herbal medicine at the Kwame Nkrumah University of Science and Technology in Ghana. He is a strong advocate for policy reform and research. His interests are in pharmacology and drug discovery.

Francis Tetteh is a medical herbalist intern at the Centre for Plant Medicine Research. He is result-driven, with vast interest in clinical research, pharmacology, and phytochemistry. He is an advocate for the safe and effective use of herbal medicine.

Bernard Kofi Turkson, is a medical herbalist, pharmacognosist, researcher in clinical trials of natural products, and lecturer at the Department of Herbal Medicine, Kwame Nkrumah University of Science and Technology, Faculty of Pharmacy and Pharmaceutical Sciences, Kumasi, Ghana.

FOREWORD

One of the biggest lessons learnt in Africa during the COVID-19 pandemic was the futility of abdicating health and socioeconomic responsibilities to external stakeholders. Empirical evidence suggests that up to two-thirds of our people use herbal or phytomedicines for their healthcare needs. Additionally, the history of drug discovery and development attests to the immense potential of thousands of plants found in diverse locations across the African continent. It is therefore heart breaking that little concerted Pan-African effort has been made to prioritise this sector and harness its great prospects to not only improve access to health care but also greatly stimulate key socioeconomic indices, such as job creation, capacity building, revenue generation, and technology transfer.

This book comes at an auspicious time when the continent urgently needs reflections on these issues and appropriate interventions. Obi Adigwe and Kofi Busia, alongside their colleagues across the continent, have adopted a Pan-African perspective in exploring existing challenges to the development of phytomedicines in Africa. They have also provided a wide range of evidence-based contextual solutions to relevant health and socioeconomic issues. Together, they have leveraged their industry and practice experience to weave an intricate conceptual combination which can help Africa utilise its phytomedicinal resources to leapfrog health and socioeconomic development.

The scholarly rigour adopted in this book, as well as the many concepts, theories, and interventions explored, makes it a useful volume that will enable policymakers, scientists, academics, practitioners, and other stakeholders to better understand the field and undertake evidence-based interventions in their respective settings.

This intellectual offering, whilst providing an evidence-based theoretical and practical basis for reforms in the sector, is also a clarion call to adopt a Pan-African perspective to problem-solving on the continent. In 2021, as the COVID-19 pandemic raged on, I was privileged to headline the All-Stakeholders International Conference on COVID-19 (ASSIC-19) convened by the authors, which advocated the adoption of a multidisciplinary Afrocentric approach to solution finding. During ASSIC 19, I gained first hand insights into the prodigious potential of the sector and the critical interventions being undertaken to harness them. With the right policies, the development of African phytomedicines can stimulate an exponential increase in access to medicines on the continent and spur socioeconomic development.

Never before has the need for Africa to look inwards for health care and socioeconomic solutions been as acute as it is at present. This is exactly what makes this book so timely. I therefore strongly recommend it to all stakeholders and individuals interested in African health care and socioeconomic development.

Professor Yemi Osinbajo, GCON, SAN
Vice President of the Federal Republic of Nigeria

PREFACE

Africa's health indices across various relevant areas have remained a source of concern, despite evidence suggesting significant ethno-pharmaceutical potential inherent in thousands of plants across the continent. For centuries, our forebears have handed down valuable knowledge from generation to generation, and even currently, two-thirds of our people still resort to phytomedicines for their healthcare needs. Ironically, despite this significant potential, there is little evidence that these resources are being harnessed in a manner that provides commensurate value for the continent and its people.

Over the past few years, Dr Busia and I have led a strident advocacy for the adoption of a Pan-African approach towards building local capacity in the utilisation of natural resources to address continental issues. We have individually and collectively argued for a smart integrated approach that builds and then leverages local capacity to harness our natural biodiversity using the highest-quality evidence-based research, development, and production techniques. Following this approach, Africa will definitely and sustainably confront its healthcare needs whilst also addressing pressing socioeconomic issues such as job creation, knowledge transfer, and revenue generation. Therefore, *A Contextual Exploration of Phytomedicines' Exploration Development in Africa* comes at an auspicious time when the continent needs these interventions most.

For this project, a deliberate effort was made to identify and co-opt practitioners, researchers, academics, and other relevant industry leaders that represent Africa's best in various thematic areas along the entire phytomedicines-medicines security value chain. In our bid to deepen relevant concepts and interrogate conceptual practices, a holistic yet multidisciplinary design was adopted to ensure a robust engagement with distinct yet connected themes such as research, production, enterprise, conservation, ecology, health policy, regulation, and critical socioeconomic factors. We also actively and deliberately sought out highly qualified colleagues and research groups from Congo, Ghana, Nigeria, Senegal, Sierra Leone, Tanzania, Zimbabwe, and other parts of the continent to ensure the adoption of a Pan-African perspective. Contributions from Europe and Asia which met the threshold were also included to ensure that the publication leveraged cutting-edge contributions underpinned by best international practices.

In the text, a strong case is made for the prioritisation of critical aspects of the phytomedicinal value chain as a catalyst for quality healthcare provision and expedited socioeconomic growth. A contextual exploration of research and development initiatives geared towards harnessing phytomedicinal potential for the achievement of universal health coverage within the African setting is also undertaken. Other critical factors which support the mainstreaming of the sector within the context of contemporary healthcare provision and acceptability amongst the citizenry are robustly examined. The nexus between strategic public health strategies such as health promotion and mass sensitisation and intellectual property rights, alongside other attributes that support successful enterprise, are reviewed.

We also explore the environmental and ecological linkages in the entire phytomedicinal cultivation and production value chain in a bid to encourage environment-friendly climate protective measures that can be embedded in various contextual processes. The book identifies various key stakeholders, including practitioners, policymakers, and regulators who all play distinct yet synergistic roles in catalysing growth and development in the sector. Sectoral opportunities and potentials are also signposted to enable stakeholders with commercial interest, such as entrepreneurs, investors, and industrialists, to gain a better understanding of this emergent sector.

The wide-ranging yet interconnected evidence-based contextual theories and solutions not only represent a novel approach in the literature but also provide a cornucopia of useful tools for various actors and stakeholders. The unique approach adopted means that all participants in the value chain, including practitioners, scientists, academics, policymakers, entrepreneurs, and investors, will be better enabled to engage and address relevant health and socioeconomic issues. The appropriate theorisation of these emergent factors will enable expedited application across Africa's varied settings and consequently enable the continent leapfrog its challenges towards developing sustainable contextual solutions to its problems. Similarly, the deliberate practical illustrations of various aspects of research, development, and production processes serve as a guide for quick replication across similar contextual settings across the continent. The concepts, theories, and interventions explored in the book will also greatly enable academics, practitioners, and policymakers to better understand the field and undertake evidence-based interventions in their respective settings.

The timing of the publication is also of paramount importance. A paradigm shift in continental policies such as the Africa Continental Free Trade Agreement (AfCTA), which encourages intra-continental trade, is a critical indicator of the catalytic role that this sector can play for widespread growth and development across the continent. The continent's lack of capacity to produce lifesaving interventions during the COVID-19 pandemic, as well as the abysmal treatment that disenfranchised it in terms of access to pharmaceuticals, also makes a strong case for this intervention. There is now a strong resolve to develop sectoral value chains that are underpinned by the continent's phytomedicinal resources. Several multibillion-dollar initiatives such as the African Development Bank (AFDB) intervention in the pharma sector have now been initiated. There is, therefore, no more critical a time to gain a better understanding of the key sectoral linkages and the evidence-based strategies that can unlock these inherent potentials. Unpacking the various linkages that the phytomedicinal sector has with policy, government, finance, regulation, and other potentially catalytic factors is exactly what this book does.

The phytomedicinal sector is suitably positioned for exponential growth, which will no doubt translate to increased access to health care on the continent as well as socioeconomic development in various relevant sectors. The multidisciplinary approach advocated by this book will enable the appropriate articulation of policies that will increase access to safe and high-quality health care and enable the realization of the great but largely untapped potential of the sector to contribute to the economy of the continent. With the appropriate synergy, the sector will not only catalyse the achievement of universal health coverage but also increase critical socioeconomic indices, such as foreign direct investment, employment generation, knowledge transfer, capacity building, and backward integration in ancillary industries.

Alongside my colleagues, it is important to appreciate the various parties that contributed directly and indirectly to the advocacy, conceptualisation, and articulation that led to the emergence of this book. Although the contributors are too many to enumerate, worthy of specific mention are the practitioners, policymakers, and stakeholders whose participation enabled anecdotal and empirical data collection and analysis that underpinned this work. Together, we have set the narrative for more a more vigorous engagement to robustly interrogate the most effective pathway to using our immense resources to exponentially improve our people's health, well-being, and lives in general.

Dr Obi Peter Adigwe
CEO/DG, NIPRD, Nigeria

ANNEX 1: LIST OF ABBREVIATIONS/ACRONYMS

ACCT	Agence de Cooperation Culturelle et Technique
ARVs	Antiretrovirals
ARIPO	African Industrial Property Organisation
CEMETRA	Experimental Centre for Traditional Medicine
COE	Centre of excellence
ECOWAS	Economic Community of West Africa States
EMRO	East Mediterranean Regional Office
HIV/AIDS	Human Immuno-Deficiency Virus/Acquired Immuno-Deficiency Syndrome
IMRA	Malagasy Institute for Applied Research
IP	Intellectual property
IPR	Intellectual property rights
KEMRI	Kenya Medical Research Institute
MAPs	Monographs on medicinal and aromatic plants
NAPRECA	Natural Products Research Network for Eastern and Central Africa
NGO	Nongovernmental organisation
NIMR	National Institute for Medicine Research
NMRA	National Medicines Regulatory Authority
NIPRD	National Institute of Pharmaceutical Research and Development

OAPI	African Intellectual Property Organisation
OAU	Organisation of African Unity
OAU/STRC	Organisation of African Unity/Scientific and Technical Research Commission
PHC	Primary health care
PROMETRA	Promotion des médicaments traditionnels améliorés/ Promotion of improved traditional medicines
R and D	Research and development
SDGs	Sustainable Development Goals
SADC	Southern African Development Community
STD	Sexually transmitted disease
TAWG	Tanga AIDS Working Group
THETA	Traditional and Modern Health Practitioners Together against AIDS
THPs	Traditional health practitioners
THPC	Traditional Health Practitioners Council
TM	Traditional medicine
TMK	Traditional medical knowledge
UHC	Universal health coverage
UNIDO	United National International Development Organization
WHA	World Health Assembly
WAHO	West African Health Organization
WHO	World Health Organization
WHO/AFRO	World Health Organisation Regional Office for Africa
WHO/HQ	World Health Organisation headquarters
WIPO	World Intellectual Property Organisation

1

Prioritisation of Phytomedicines as a Stimulus for Increasing Access to Quality Healthcare for Improved Socioeconomic Development

Mavis Boakye-Yiadom[1], Doris Kumadoh[2,3], Zhou Jephias Redemptor[1], Francis Tetteh[1], Irene Aasam Aabeinir[1]

[1]Clinical Research Department, Centre for Plant Medicine Research, Mampong-Akuapem, Ghana
[2]Department of Pharmaceutics and Quality Control, Centre for Plant Medicine Research, Mampong-Akuapem, Ghana
[3]Production Department, Centre for Plant Medicine Research, Mampong-Akuapem, Ghana

1.1 Introduction

African traditional medicine (ATM) is complex, encompassing the practices of divination, spirituality, and herbal medicine usage (Ozioma *et al.*, 2019). Herbal medicine, which is common to all forms of traditional medicine (TM) practices, is the core of African traditional medicine (Ozioma *et al.*, 2019; WHO Regional Office for South-East Asia, 2010). Contrary to the supposed 'complementary' role of TM, various reports indicate that 80% of people in sub-Saharan Africa, solely depend on it (Shewamene *et al.*, 2017; Nyame *et al.*, 2021). The shifting belief system, where Christianity hugely supplanted indigenous traditional religions (Berends, 1993; Sugishita, 2009), and the colonial legacy seem to have inveterate Western medicine in the African formal healthcare system. However, despite the mounting evidence of the therapeutic benefits and widespread use of medicinal plants, there seems to be some disconnect between the health priorities of governments in Africa and the healthcare needs of the citizenry as herbal medicines rarely feature in national health plans and strategies. The widely reported surge in herbal medicine usage (Eisenberg *et al.*, 1997; Roy-Byrne *et al.*, 2005; Calapai and Caputi, 2007; Bernstein *et al.*, 2021) may be an inaccurate narrative in the African context, or rather it describes a total global trend, particularly in the Western world; the surge occurring in Africa is in public discussions, scientific scrutiny, and dialogue.

In an attempt to categorise and subcategorise traditional medicine practice, various terms have been used to describe it. These include complementary and alternative medicine (CAM), unorthodox medicine (UM), and traditional and alternative medicine (TAM). In Africa, the term 'African traditional medicine' is often used interchangeably with these terms.

For the purpose of this article, the authors prefer to use the term 'herbal medicine' or 'phytomedicine' but may from time to time introduce TM in the discussion.

The authors take phytomedicine, herbal medicine, or plant medicine to mean the same and define it as any therapeutic substance whose origin can be traced to plants and whose pharmacological properties have not been artificially enhanced. By their nature, phytomedicines consist of a large number of active constituents such as flavonoids, phenols, and alkaloids, in contrast to drugs that usually consist of a single recognisable ingredient. However, the active constituents in phytomedicines can be isolated and characterised and the metabolites responsible for a particular effect identified.

The scope of what can be considered phytomedicines is contentious, with the evolution of various interesting terms to categorise them. The USA Dietary Supplement Health and Education Act of 1994 categorised vitamins, minerals, amino acids, herbs, and other botanicals as nutrient supplements. Terms such as 'nutraceuticals', which show a strong resemblance to what can be termed 'phytomedicine', have also risen. However, there is no homogeneous and dominant African view on this issue as terminologies and policies are usually takedowns or borrowed from the Western world. It is interesting and encouraging to note that sometimes phytomedicines are regulated as prescription or non-prescription medicines and are traded with health claims (WHO Global Report on Traditional and Complementary Medicine, 2019).

The current authors deprecate the idea that 'herbs are food' and support the notion to depart from it as it is a propaganda that may fuel toxicity because of misuse (Zhang *et al.*, 2015) or promote non-expert prescriptions and undervalues the therapeutic benefits of medicinal herbs.

1.3. Prioritisation of phytomedicines

1.3.1 The need for prioritisation

As mentioned earlier, the problems stunting the growth of the herbal medicine industry have less to do with the users of phytomedicines and more to do with the indifference of the policymakers. Prioritisation of herbal medicines, therefore, requires official recognition by policymakers that these medicines can be at par with or better than Western medicines as medication. This will require courage and firm decision-making as African governments may have to ration their meagre financial resources to include herbal medicines, probably to the displeasure of international agencies that are usually major partners in health care. Countries in Asia have managed this impressively as shown in the 13.6 million traditional Chinese medicine inpatients in 2009 (Qi, 2013). The major issue is that much of the healthcare systems in sub-Saharan Africa (SSA) have inadequate government funding, characterised by a huge dependence on external (donor) funding (Asante *et al.*, 2020) and largely handled by the private sector (International Finance Corporation, 2008; 2011). Buckling under the weight of the COVID-19 pandemic, African governments face a dilemma as there is a dual public health and economic crisis whose ripple effects may last for years (Union *et al.*, 2020). Therefore, there is a higher likelihood that governments will concentrate their efforts on rebuilding their economies, thus contracting their health budgets, culminating in an increase in dependence on donors to finance health care. The herbal medicine industry can be harnessed for the greater good as medicines are generally affordable (Wachtel-Galor and Benzie, 2011), hence requiring less monetary investments as the expertise is already ubiquitous in communities. The next level will be to assess the safety and efficacy scientifically.

1.3.2 Science as a tool: Empowering the scientist

Proponents of phytomedicines hail them for their perceived effectiveness, especially where orthodox medicines fail (Wachtel-Galor and Benzie, 2011). The safety of these medicines has, however, been heavily questioned (Aydin et al., 2016; Chanda et al., 2015), with the growing calls for evidence-based medicine, which is backed by credible data.

Knowledge on herbal medicines has transcended generations through oral tradition. This has enabled people who may not be necessarily traditional healers to self-medicate or medicate others. Thus, scientists with interest in medicinal plants only have to reach out to these communities to extract this invaluable knowledge. The heterogeneous nature of the African flora presents an opportunity for collaboration. Much of the evaluation of these plants has been possible in animal studies with much ground needed in clinical trials. Apart from these, assessing the active constituents of the plant to predict its activity will also help. However, it is arguable whether phytomedicines, with their extensive history of usage since ancient times and outspoken evidence of efficacy through empirical observation and trial and error (Karunamoorthi et al., 2013), do really need to be subjected to the same standards as drugs or even whether these prescribed standards may be wholly applicable (Calapai and Caputi, 2007) or a different standard has to be designed for them.

1.3.3 A science that is relevant to the ordinary person

Recently, there has been a plethora of scientific investigations to identify the active constituents in plants and link them to their ethnomedical uses. Various pharmacological studies have also been carried out on a number of models to further buttress the evidence for the use of herbal medicines. Scientific publications in reputable journals are consistently churned out, with promising information on the therapeutic benefits of phytomedicines. While this is a welcome development in the journey towards promoting herbal medicines as dependable pharmacological agents, there are still calls for comprehensive clinical trials to validate their ethnomedical uses. This is not impossible, as has been shown by the drug Veregen (Polyphenon E ointment), the first prescription botanical (herbal) medicine to be approved by the US Food and Drug Administration (Bhusnure *et al.*, 2019) as its safety was verified in randomised, double-blind (placebo)-controlled clinical trials (Stockfleth *et al.*, 2008; Hoy, 2012).

In Africa, the Ghanaian herbal medicine industry has exhibited some degree of advancement, with plants being developed into products, which are well packaged and approved by the Food and Drugs Authority (FDA), with prior certification by research centres such as the Centre for Plant Medicine Research, where they are investigated for their efficacy and safety (Aziato and Antwi, 2016). Additionally, in a survey on the *Capacity for Clinical Research on Herbal Medicines in Africa*, Willcox et al. (2012) realised a strong interest among African researchers to conduct clinical trials although they are hampered by various constraints. In that survey, twenty-three trials had been done, with only five publications and fifty-four possible trials in the future. These examples give some evidence that herbal medicines can pass the rigorous scientific processes applied to drugs. But it is important that these medicines are patented to protect the owners of the knowledge.

1.3.4 Patents and rewards

Protection of traditional medical knowledge has for long been a subject of debate. During the COVID-19 pandemic, the issue of patents became topical, especially with the discovery and manufacture of anti-COVID-19 vaccines. The blatant reluctance by vaccine manufacturers to waive the patency was based on the argument that it was unfair to those who had sacrificed their resources. This goes to show the extent to which patents and rewards can stimulate interest in research and development among scientists. With protections such as *sui generis*, pharmaceutical manufacturing companies and scientists may not be willing to invest their resources in research as the rewards may not be sufficient or even be unavailable. Thus, in prioritising phytomedicines, there is a need to re-evaluate patency laws to strike a balance between protecting communities from exploitation and encouraging local scientists and their funding bodies to venture into research and development. The vastness of flora presents a fertile ground for discovery of very potent compounds that may also act as new leads for synthesis of new compounds. There is therefore a need for an all-inclusive mindset to promote the research and development of phytomedicines.

1.3.5 Non-discrimination: An all-round agenda

As previously mentioned, policymakers in Africa have largely neglected phytomedicines in their national policies and, as a result, do not allocate any budget to them. Phytomedicines are still viewed as esoteric and with suspicion because of the paucity of scientific data and have only been featured in discussions when the demand has made them hard to ignore. To prioritise phytomedicines, there is an urgent need to remove the discrimination and stigma associated with them. To do this, there have been some suggestions to change the term 'traditional medicine' into something more 'contemporary and acceptable', such as 'phytotherapy', 'natural medicine', or even 'botanical medicine'.

1.4.0 Harnessing phytomedicines for quality health care

Quality health care involves getting access to the right diagnosis and treatment protocols in a prejudice-free manner in the event of illness. In Africa, herbal medicine practice is left mostly to the informal sector where safety practices and appropriate technology leave much to be desired. The apparent neglect of herbal medicines and practices in Africa has dire consequences, as in many cases, issues of safety are at the discretion of the manufacturer. For instance, in Ghana, a lot of negative reportage about herbal medicines has taken over the airwaves as a means of cautioning consumers on the potential hazards of herbal medicines. But dissuading people from consuming herbal medicines may be a futile exercise as these medicines are inveterate in primary health care for their perceived efficacy (Burton et al., 2015; Mokgobi, 2013).

Although governments seem to have a lukewarm attitude towards phytomedicines, the WHO (1978) recognises them as one of the surest means to achieve universal health coverage of the world's population. This is because its records have consistently shown a very high patronage, with about 70% of Africans in particular, depending on them for their healthcare needs. Therefore, at the 1978 WHO-sponsored International Conference on Primary Healthcare at Alma-Ata (USSR), governments were urged to integrate traditional medicine into their formal healthcare systems to effectively improve access to quality and affordable health care.

Primary health care (PHC) is key to the development of a national health policy. As noted in the Alma-Ata Declaration of 1978, it is an essential health care based on practical, scientifically sound, and socially acceptable methods and technology made universally acceptable to individuals and families in the community through full participation, and at an affordable cost to the community and the country, to maintain at every stage of their development in the spirit of self-reliance and self-determination (White, 2015). PHC is the first level of contact for the individual, family, and the community within the national healthcare system, bringing health care as close as possible to where people live and work, and thus constitutes the first element of a continuing healthcare process (WHO, 1978a). A health system based on primary health care was adopted as the means of achieving the goal of health for all by the year 2000. Hence, the goal of 'health for all' remains unattained in all such countries. It is interesting to note the dual failure of governments to achieve 'health for all' and integrate phytomedicines into the formal health care. It is debatable that these failures are not mutually exclusive. Thus, integration remains a viable option and a 'must' vehicle in the drive towards achieving 'health for all'.

In Ghana, integration by the policy of the Ministry of Health (MOH) is the incorporation of herbal medicine into the mainstream healthcare delivery system at an existing health facility (Good, 1977). This involves the introduction of herbal medicines, techniques, and knowledge of herbs into the country's mainstream healthcare delivery system.

Examining the philosophy from the critical viewpoint of the definition of primary health care, it is easy to assess orthodox medicine practice alongside the traditional type of health care in the African context. Specifically, in the areas of social acceptability, cost affordability, self-reliance, cultural compatibility, relevance, and community participation, orthodox or modern/Western-based medicine has not been adequate for the majority of African populations, and if progress is to be made, there is an inevitable need for official integration of phytomedicines and the utilisation of herbal medical practitioners into the PHC system in Africa (WHO Geneva, 2002).

According to the most recent WHO strategy on traditional medicine (TM) 2014–2023, the ratio of traditional healers to the population in Africa is 1:500, whereas the ratio of medical doctors to the population is 1:40,000 (Qi, 2013). Hence, access to the Western type of health institutions is out of reach of most people in terms of distance and costs, especially at the village setting. On the other hand, so long as every rural setting in Africa cannot yet be provided with basic health care, readily available and affordable drugs, orthodox medicine, as currently made available today in most African countries, cannot dispense with the support of phytomedicine practitioners. Therefore, the most workable health agenda for Africa is the institutionalisation of phytomedicines. Hamilton (2004) posits that to fully tap into traditional medicines, another approach is their official recognition to operate in parallel (not in fusion) with orthodox medicine. Such a system would liberalise traditional medicine practice, allowing it not to constrict by trying to imitate conventional medicine, but to have its own independent evolution, but guided by modern scientific methods and practices. However, the downside may be that an independent system would need to start from zero, building infrastructure and equipment which would rather be shared in an integrated system. The hope in this happening would be far-fetched, especially considering the financial difficulties already being faced in healthcare systems making an effort for integration. Effective health agenda for the African continent can never be achieved by orthodox medicine alone unless it is complemented by traditional medicine practice (Pandey, 2013).

Some of the attributes of herbal medicines which qualify it as an essential contributor to the quest for attaining quality health care include the following:

1.4.1 Accessibility: Phytomedicines predate modern pharmaceutical and medical systems. In the African context, plants close to people's homes are the immediate port of call for individuals in the event of illness. The only impediment to the use of phytomedicines is the understanding of the art/science behind their activities (phytotherapy) (Patel et al., 2012). It has been understood as established by Paracelsus that 'everything is a poison; it is only the dose that differentiates a poison from remedy'. In view of this, it will be necessary to educate people about the rational use of phytomedicines for optimal health, and this can go a long way to prevent harm from improper use, which will be a crucial contribution to the goal to achieve quality health care.

1.4.2 Affordability: Most Africans, particularly rural dwellers, cannot afford orthodox-based primary health care because of financial challenges (Shewamene et al., 2017). Since the cost of producing most dosage forms of herbal medicine is usually lesser than that of conventional medicine, phytomedicines become more affordable to society, and this improves their access to quality health care (Maroyi, 2013).

1.4.3 Safety: Majority of herbal remedies have been shown to have wide therapeutic indices; hence, they could be taken in relatively larger quantities above their recommended doses before they become lethal. Unlike herbals, which have multiple components, most orthodox medicines, being isolated single, active constituents, are prone to lots of side effects and contraindications because of their specificity to receptors (Sam, 2019; Usmani et al., 2021). However, vigilance is encouraged as studies have shown that the abuse of herbal medicines can lead to severe toxic effects (Zhang et al., 2015).

1.4.4 Lead compounds for essential medicines: It is estimated that 100 natural product-based drugs are in clinical studies, and of the 252 drugs on the World Health Organization's (WHO) Essential Medicines List, 11% are exclusively derived from plants (Wachtel-Galor and Benzie, 2011). Furthermore, majority of conventional medicines in use today were first discovered from plants and then isolated and reproduced in larger quantities with special modifications to achieve desired results, such as short onset of actions and sustained release among others (Qazi Majaz et al., 2016). It is estimated that 25% of medicines such as aspirin, artemisinin, ephedrine, and paclitaxel were derived from plants (Zhang et al., 2015). This, therefore, calls for more research into the use of phytomedicines to attain quality health care.

1.4.5 Better option for the treatment of chronic diseases: Reports from various countries have pointed to a high prevalence of use and trust in herbal medicines, particularly for the treatment of chronic diseases (Issa and Basheti, 2017; Peltzer and Pengpid, 2019). For example, herbal medicines have shown great potential in treating burdensome conditions such as tuberculosis, human immunodeficiency virus, and cancers (Khan and Ahmed, 2019).

Although conventional medicines have faster onset of action with controlled duration due to structural modifications, phytomedicines, which have multiple phytochemical constituents working in a complementary and synergistic manner (Karimi et al., 2015), are often able to provide curative effects as compared to the former, which basically alleviate or suppress symptoms (Barnes et al., 2007). However, some plants such as *Digitalis purpurea* have been shown to have actions closer to that of pharmaceuticals (Karimi et al., 2015).

1.5.0 Jobs and revenue generation

The herbal medicine industry is rarely or never mentioned as an official means of generating revenue in Africa. However, the figures quoted by a few market surveys are quite revealing and should engage the attention of policymakers. For example, the global herbal medicine industry is estimated at $411.2 billion by the year 2026 (Market Study Report, 2020). Some figures quoted as generated by the herbal medicine industries in some African countries are shown in Table 1. Figure 1 also shows an expected global increase in herbal products consumption.

Table 1: Figures quoted as generated by the herbal medicine industries in some African countries

Country	Year	Amount	Reference
South Africa	2007	2.9 billion rand	(Mander et al., 2007)
Ghana	2012	US$7.8 million	(Van Andel et al., 2012)
Benin	2014	US$2.7 million	(Quiroz et al., 2014)

Figure 1: Global wax in revenue and consumption of plant-based medicine (Jibril et al., 2019)

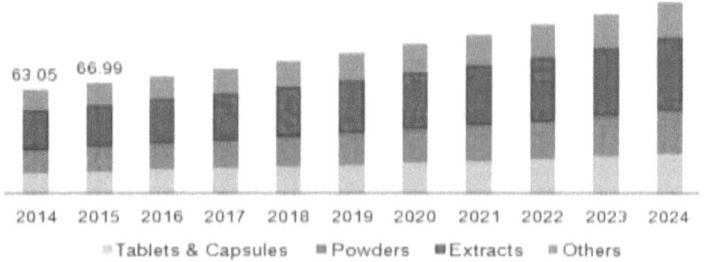

Figure 1. Global herbal medicine market revenue, by product, 2014–2024 (USD Billion)
(source: HEXA RESEARCH, Herbal Medicine..., 2017)

The above graph shows the magnitude of the industry and its potential for massive revenue generation if harnessed properly. In Africa, much of herbal medicine is handled by traditional healers, individuals, and informal traders who elude governments or official statistics. Poor households are able to eke a living out of medicinal plant trading (Rasethe et al., 2019), although the lack of sophistication and innovation of herbal products may impede rewards. Many of the recorded statistics are for exports, and it is likely that monetary values involved in domestic transactions could be much higher. Thus, this apparent omission is a travesty, and the remedy is to consider the herbal option as a fiscal revenue source. Additionally, positive government attitude towards herbal medicines will boost confidence and create new allies in the private sector, which may be willing to venture and invest in this unexplored medical industry. With governments showing confidence in the industry, doubts about the phytomedicine industry being an avenue worth exploring will be dispelled.

The following are some of the ways phytomedicines can be used to generate jobs and revenue.

1.5.1 Trade and exportation

The African continent has a heterogeneous climate and geography, with varied vegetation. This gives an opportunity to trade in medicinal plants between countries, which may be both labour and capital intensive (Mandal, 2020). Some of the plants grow naturally in hard-to-reach terrains, which only the local people know very well. Engaging these people to do the collection and transportation will be essential in creating jobs. Once trade is involved, professionals such as accountants can be roped in to manage cash flows.

1.5.2 Medicinal plant farming

There is no doubt that climate change poses an existential threat to rare and endangered plant species. Hamilton (2004) suggested four distinct subsystems involved in medicinal plant conservation, namely, (1) production systems and in situ conservation; (2) commercial systems; (3) ex situ conservation, propagation, domestication, and the breeding of crop varieties; and (4) new product discovery. This shows the complexities involved in the processes of medicinal plant conservation. Besides the threat of extinction, changes in climatic conditions can also affect the level of bioactive constituents in medicinal plants (Applequist et al., 2020) and render them ineffective or even toxic. In addition, the loss of vegetation means that some knowledge and healing practices are also lost (Malan et al., 2015). The practice of herbal medicine has largely been primitive, where people collect them from the wild (Street et al., 2008). This has left the practice at the mercy of natural factors such as the abundance of rains or lack of it, soil fertility, and other environmental factors which significantly compromise the quality of medicinal plants (Yang et al., 2018). As plants may be collected from the wild, sometimes hard terrains, locals with limited knowledge in sustainable harvesting are employed (Rasethe et al., 2019). Additionally, improper collection procedures, processing, and warehousing leading to the product adversely affect the international market value of African natural plant products (Street et al., 2008). Medicinal plant farming is a sure way of preventing the extinction of some plant species and the loss of their associated therapeutic uses. Apart from that, farming ensures sustainability and dependability. Medicinal plant farming ensures that environmental conditions are optimised and good manufacturing practices are adhered to, which significantly improves the quality of medicinal plants and thus the products. Moreover, high-yield and high-quality medicinal plants can be harvested from artificially controlled environmental conditions such as greenhouses and bionic cultivation, which also allow plant development, growth, and stable production (Yuan et al., 2020). Controlled environments can also allow research to be carried out on these plants. Additionally, Hamilton (2004) notes that the best way of conservation of plants involves systems' strictness on conservation or sustainable production (or both) at the sites of medicinal plant growth. Thus, in an artificial system, sustainable harvesting can be prioritised as well. People with different skill sets such as botanists, agronomists, and even economists could be employed. Apart from that, the ongoing discussion on the efficacy of medicinal cannabis in treating certain medical conditions (Amin and Ali, 2019; Breijyeh et al., 2021; Slawek et al., 2022), can be an opportunity for the African continent to exploit (Owusu et al., 2021; Gerwel, 2018) to position itself as a leader in cannabis export.

1.5.3 The manufacturing industry

The pharmaceutical industry in Africa is struggling to mark its presence. At the moment, competing with big pharma in producing Western medicines is not a commercially viable option. However, phytomedicines are being produced haphazardly in poor environs, often on a small scale. Their quality is compromised, and thus, they have a lower market value, which makes their contribution to the economy negligible (Street et al., 2008). With the boom in the herbal medicine industry, improving it has the potential to make Africa a global leader in the sector and create various jobs to boost the economy by generating valuable revenue.

1.5.4 Intergenerational partnership

The elderly, usually uneducated, are the custodians of traditional medical knowledge in Africa. Although they can rightly claim monopoly over that knowledge, young people can take the stage when it comes to yoking modern-day technology to revolutionise and secure the knowledge for the future. Ensuring that knowledge on plant medicine is passed on to the next generations depends on whether the young of today will take special interest in these plants.

Previously, it might have been easier for knowledge to transcend generations as traditional medicine was the only form of medicine, and thus, communities were compelled to learn. However, in today's Africa, Western medicine has almost supplanted traditional medicine, and given that scientists may not be able to document ethnobotanical knowledge at a fast enough pace, the youth of today has a duty to jealously guard and preserve traditional medical knowledge to ensure that it does not die with its custodians. Thus, a hand-in-glove relationship is essential for reaping the rewards in the phytomedicine industry. Reports have pointed to the unwillingness of some communities to divulge their folkloric knowledge and cases of foreign and even local scientists getting credit for local traditional medical knowledge at the expense of the local community. The elderly can be paid for divulging their traditional medicine secrets. They can also be given little education, which will allow them to be employed as consultants while getting paid for their roles.

1.6.0 How phytomedicines can promote African Continental Free Trade Area (AfCFTA)

The African Continental Free Trade Area (AfCFTA) is an Afro-optimism strategy fuelled by the view that the economic growth of Africa can be stimulated from within through unity. It can be argued that it is a remnant of the original ideologies that birthed the Organisation of African Unity (OAU) (Moorosi Leshoele, 2020). The AfCFTA began on 1 January 2021, with the aim of achieving economic growth through the promotion of trade between its members, thereby creating a single market for goods and services (Tsowou and Davis, 2021). This a wholesome strategy which, when properly exploited, has the potential to impact each and every sector in Africa.

The phytomedicine industry can contribute to the success of AfCFTA by exploring and exploiting the opportunities that it brings. The continent is already known for its high disease burden because of treatable infections and for being a beggar of medical supplies. This unfortunate situation is a ready market if African governments partner to authorise trade in phytomedicines. Additionally, it is an opportunity to collaborate in creating standards, specifically for herbal medicines, as well as congruent policies and regulations on the practice of herbal medicines. Congruent standards, policies, and regulations are essential as they can allow ease in trade of products as regulations will not be an impediment. As mentioned earlier, traditional medicine practice is influenced by culture and thus differs from one African country to another. For example, *Cryptolepis sanguinolenta* from Ghana is a plant renowned for treating malaria, yet it is not found in Southern Africa, which has its own famous plant, *Agathosma betulina* (Berg.) known for its diuretic properties (Street and Prinsloo, 2013). This presents an opportunity for collaboration as useful plants can be exported from one African country to another with ease, broadening the consumer base. In the AfCFTA, it would be cheaper for an African country to import medicinal plants compared to importing orthodox drugs from the Western world. Furthermore, this creates a platform for collaboration, innovation, and development of herbal products. As the market increases, the population study sample is also made available to provide data for research purposes.

1.7.0 Conclusion

From the foregoing discussion, it is clear that phytomedicines have not lost their prominence in the lives of human beings. The quest for potent synthetic drugs has been at the expense of phytomedicines. In history, the developed world experienced a shift from herbal medicines to orthodox medicines and has recently been experiencing a refocus on herbal medicines. On the other hand, the developing world has largely depended on herbal medicines for their populations' healthcare needs, although public policy and media have swayed public opinion to think otherwise. This discussion highlights that phytomedicines have an even greater role to play in both health care and the economy of African countries. The major challenge facing the practice of herbal medicine is that it is still locked in the past, and this has limited its standing in society. This discussion advocates for a renewal in how herbal medicines are viewed and treated. The herbal medicine practice can be elucidated, evaluated scientifically, developed using modern technology, and harnessed as a driver in economic development. However, it will take more than the usual political rhetoric and dishonesty, but concrete action and sheer determination on the part of governments and the corporate world.

References

1. Amin MR, Ali DW (2019). Pharmacology of Medical Cannabis. Advances in Experimental Medicine and Biology, *1162*, 151–165. https://doi.org/10.1007/978-3-030-21737-2_8
2. Applequist WL, Brinckmann JA, Cunningham AB, Hart RE, Heinrich M et al (2020). Scientists' Warning on Climate Change and Medicinal Plants. Planta Medica, 86(1), 10–18. https://doi.org/10.1055/a-1041-3406
3. Asante A, Wasike WSK, Ataguba JE (2020). Health Financing in sub-Saharan Africa: From Analytical Frameworks to Empirical Evaluation. Applied Health Economics and Health Policy, 18(6), 743–746. https://doi.org/10.1007/s40258-020-00618-0
4. Aydin A, Aktay G, Yesilada E (2016). A guidance manual for the toxicity assessment of traditional herbal medicines. Natural Product Communications, 11(11), 1763–1773. https://doi.org/10.1177/1934578x1601101131
5. Aziato L, Antwi HO (2016). Facilitators and barriers of herbal medicine use in Accra, Ghana: an inductive exploratory study. BMC Complementary Medicine and Therapies 16, 142. https://doi.org/10.1186/s12906-016-1124-y
6. Barnes PM, Bloom B (2007). Complementary and alternative medicine use among adults and children: United States, 2007. National Health Statistics Reports, 2008 Contract No.: 12
7. Berends, W. (1993). African Traditional Healing Practices and the Christian Community. Missiology, 21(3), 275–288. https://doi.org/10.1177/009182969302100301

8. Bernstein N, Akram M, Yaniv-Bachrach Z, Daniyal M (2021). Is it safe to consume traditional medicinal plants during pregnancy?. Phytotherapy Research: PTR, 35(4), 1908–1924. https://doi.org/10.1002/ptr.6935

9. Bhusnure OG, Shinde MC, Vijayendra SS, Gholve SB et al (2019). Phytopharmaceuticals: An emerging platform for innovation and development of new drugs from botanicals. Journal of Drug Delivery and Therapeutics, 9(3-s), 1046–1057. https://doi.org/10.22270/jddt.v9i3-s.2940

10. Breijyeh Z, Jubeh B, Bufo SA, Karaman R et al (2021). Cannabis: A Toxin-Producing Plant with Potential Therapeutic Uses. Toxins, 13(2), 117. https://doi.org/10.3390/toxins13020117

11. Burton A, Smith M, Falkenberg T. (2015). Building WHO's global Strategy for Traditional Medicine. European Journal of Integrative Medicine, 7(1), 13–15. https://doi.org/10.1016/J.EUJIM.2014.12.007

12. Calapai G, Caputi AP (2007). Herbal medicines: can we do without pharmacologist? Evidence-Based Complementary and Alternative Medicine: eCAM, 4(Suppl 1), 41–43. https://doi.org/10.1093/ecam/nem095

13. Chanda S, Parekh J, Vaghasiya Y, Dave R et al (2015). Medicinal Plants -From Traditional Use to Toxicity Assessment: a Review. International Journal of Pharmaceutical Sciences and Research IJPSR, 6(7), 2652–2670. https://doi.org/10.13040/IJPSR.0975-8232.6(7).2652-70

14. Eisenberg DM, Davis RB, Ettner SL, et al. (1997). Trends in Alternative Medicine Use in the United States, 1990–1997: Results of a Follow-up National Survey. *JAMA*. 1998;280(18):1569–1575. doi:10.1001/jama.280.18.1569

15. Gerwel H (2018). Institutional innovation and pro-poor agricultural growth: cannabis cultivation in the Eastern Cape province of South Africa as fertile opportunity.

16. Good CM (1977). Traditional medicine: An agenda for medical geography. Social Science and Medicine (1967), 11(14–16), 705–713. doi: 10.1016/0037-7856(77)90156-1

17. Hamilton AC (2004). Medicinal plants, conservation and livelihoods. Biodiversity and Conservation 13, 1477–1517.#

18. Hoy SM (2012). Polyphenon E 10% Ointment. American Journal of Clinical Dermatology 13, 275–281 (2012). https://doi.org/10.2165/11209370-000000000-00000

19. International Finance Corporation (2011). The Business of Health in Africa. 8–13. https://www.unido.org/fileadmin/user_media/Services/PSD/BEP/IFC_HealthinAfrica_Final.pdf

20. Issa RA, Basheti IA (2017). Herbal Products Use Among Chronic Patients and its Impact on Treatments Safety and Efficacy: A Clinical Survey in the Jordanian Field. Trends in Medical Research, *12: 32–44.*

21. Jibril AB, Kwarteng MA, Chovancova M (2019). A demographic analysis of consumers' preference for green products

22. Karimi A, Majlesi M, Rafieian-Kopaei M (2015). Herbal versus synthetic drugs; beliefs and facts. Journal of Nephropharmacology, 4(1), 27–30.

23. Karunamoorthi K, Jegajeevanram K, Vijayalakshmi J, Mengistie E (2013). Traditional Medicinal Plants: A Source of Phytotherapeutic Modality in Resource-Constrained Health Care Settings. Journal of Evidence-Based Complementary and Alternative Medicine, 67–74. https://doi.org/10.1177/2156587212460241

24. Mahomoodally MF (2013). Traditional Medicines in Africa: An Appraisal of Ten Potent African Medicinal Plants. Evidence-Based Complementary and Alternative Medicine, v2013, ID 617459, 14 pages, 2013. https://doi.org/10.1155/2013/617459

25. Malan DF, Neuba DFR, Kouakou KL (2015). Medicinal plants and traditional healing practices in ehotile people, around the aby lagoon (eastern littoral of Côte d'Ivoire). Journal Ethnobiology Ethnomedicine, 11, 21. https://doi.org/10.1186/s13002-015-0004-8

26. Mandal RA (2020). Revenue and Employment Generation from Medicinal Herbs in Darchula, Nepal. Open Access Journal of Biogeneric Science and Research, 1(4). https://doi.org/10.46718/jbgsr.2020.01.000023

27. Mander M, Ntuli L, Diederichs N, Mavundla K (2007). Economics of the traditional medicine trade in South Africa: health care delivery. South African Health Review, January 2007, 189–196.

28. Maroyi A (2013) Traditional use of medicinal plants in South-Central Zimbabwe: Review and perspectives. Journal of Ethnobiology and Ethnomedicine. 2013;9(31):1–18. DOI: 10.1186/1746-4269-9-31

29. Mokgobi MG (2013). Towards integration of traditional healing and western healing: Is this a remote possibility?. African Journal for Physical Health Education, Recreation, and Dance, (Suppl 1), 47–57.

30. Khan MSA, Ahmad I (2019). Herbal Medicine: Current Trends and Future Prospects,Editor(s): Mohd Sajjad Ahmad Khan, Iqbal Ahmad, Debprasad Chattopadhyay. New Look to Phytomedicine, Academic Press, 3–13, ISBN 9780128146194. https://doi.org/10.1016/B978-0-12-814619-4.

31. Nyame S, Adiibokah E, Mohammed Y *et al.* (2021). Perceptions of Ghanaian traditional health practitioners, primary health care workers, service users and caregivers regarding collaboration for mental health care. BMC Health Services Research 21, 375. https://doi.org/10.1186/s12913-021-06313-7

32. Leshoele M (2020). AfCFTA and Regional Integration in Africa: Is African Union Government a Dream Deferred or Denied? Journal of Contemporary African Studies, DOI: 10.1080/02589001.2020.1795091

33. Owusu NO, Arthur B, Aboagye EM (2021). Industrial hemp as an agricultural crop in Ghana. *J Cannabis Res* **3,** 9 (2021). https://doi.org/10.1186/s42238-021-00066-0

34. Ozioma EJ, Nwamaka Chinwe OA (2019). Herbal Medicines in African Traditional Medicine. In (Ed.), Herbal Medicine. IntechOpen. https://doi.org/10.5772/intechopen.80348

35. Pandey M, Rastogi S, Rawat A (2013). Indian traditional ayurvedic system of medicine and nutritional supplementation. Evidence-Based Complementary and Alternative Medicine. 2013;2013:1–12. DOI: 10.1155/2013/376327

36. Patel D, Kumar R, Laloo D, Hemalatha S (2012). Natural medicines from plant source used for therapy of diabetes mellitus: An overview of its pharmacological aspects. Asian Pacific Journal of Tropical Disease. 2012:239–250. DOI: 10.1016/S2222-1808(12)60054-1

37. Peltzer K, Pengpid S (2019). The use of herbal medicines among chronic disease patients in Thailand: a cross-sectional survey. Journal of Multidisciplinary Healthcare, 12, 573–582. https://doi.org/10.2147/JMDH.S212953.

38. Qazi MA, Molvi K (2016). Herbal medicine: A comprehensive review. Journal of Pharmaceutical Research. 2016;8(2):1–5

39. Qi Z (2013). The WHO traditional medicine strategy 2014–2023. Global Health History Seminar on Traditional Medicine and Ayurveda, March, 1–28.

40. Quiroz D, Towns A, Legba SI, Swier J, Brière S et al. (2014). Quantifying the domestic market in herbal medicine in Benin, West Africa. Journal of Ethnopharmacology, 151(3), 1100–1108. https://doi.org/10.1016/J.JEP.2013.12.019

41. Rasethe MT, Sebua SS, Maroyi A (209). Medicinal Plants Traded in Informal Herbal Medicine Markets of the Limpopo Province, South Africa. Evidence-Based Complementary and Alternative Medicine, 11 pages, 2019. https://doi.org/10.1155/2019/2609532

42. Roy-Byrne PP, Bystritsky A, Russo J, Craske G.M, Sherbourne DC et al. (2005). Use of Herbal Medicine in Primary Care Patients With Mood and Anxiety Disorders, Psychosomatics, 46(2): 117–122, https://doi.org/10.1176/appi.psy.46.2.117.

43. Sam S (2019). Importance and effectiveness of herbal medicines. Journal of Pharmacognosy and Phytochemistry, 8(2): 354-357.

44. Street RA, Stirk WA, Van Staden J (2008). South African traditional medicinal plant trade—Challenges in regulating quality, safety and efficacy, Journal of Ethnopharmacology, Volume 119, Issue 3, 2008, Pages 705–710, ISSN 0378-8741, https://doi.org/10.1016/j.jep.2008.06.019.

45. Shewamene Z, Dune T, Smith C.A (2017). The use of traditional medicine in maternity care among African women in Africa and the diaspora: a systematic review. BMC Complementary Alternative Medicine, 17, 382 (2017). https://doi.org/10.1186/s12906-017-1886-x

46. Slawek DE, Curtis SA, Arnsten JH, Cunningham CO (2022). Clinical Approaches to Cannabis: A Narrative Review. The Medical Clinics of North America, 106(1), 131–152. https://doi.org/10.1016/j.mcna.2021.08.004

47. Stockfleth E, Beti H, Orasan R, Grigorian F et al., (2008). Topical Polyphenon E in the treatment of external genital and perianal warts: a randomized controlled trial. The British Journal of Dermatology, 158(6), 1329–1338. https://doi.org/10.1111/j.1365-2133.2008.08520.x

48. Street RA, Prinsloo G (2013). Commercially Important Medicinal Plants of South Africa: A Review. Journal of Chemistry, 1–16.

49. Sugishita K (2009). Traditional Medicine, Biomedicine and Christianity in Modern Zambia. Africa, 79, 435–454.

50. Tsowou K, Davis Jr (2021). Reaping the AfCFTA Potential Through Well-Functioning Rules of Origin. Journal of African Trade, 8(2, Special Issue): 88 - 102

51. OECD Policy Responses to Coronavirus (COVID-19) (2020). COVID-19 in Africa : Regional socio-economic implications and policy priorities. https://www.oecd.org/coronavirus/policy-responses/covid-19-and-africa-socio-economic-implications-and-policy-responses-96e1b282/. Accessed on 8-01-2023

52. Usmani J, Khan T, Ahmad R, Sharma M (2021). Potential role of herbal medicines as a novel approach in sepsis treatment. Biomedicine and Pharmacotherapy (Biomedecine et Pharmacotherapie), 144, 112337. https://doi.org/10.1016/j.biopha.2021.112337

53. Van Andel T, Myren B, Van Onselen S (2012). Ghana's herbal market. Journal of Ethnopharmacology, 140(2), 368–378. https://doi.org/10.1016/J.JEP.2012.01.028

54. Wachtel-Galor S, Benzie IFF (2011). Herbal Medicine: An Introduction to Its History, Usage, Regulation, Current Trends, and Research Needs. In: Benzie IFF, Wachtel-Galor S, editors. Herbal Medicine: Biomolecular and Clinical Aspects. 2nd edition. Boca Raton (FL): CRC Press/Taylor & Francis; 2011. Chapter 1. Available from: https://www.ncbi.nlm.nih.gov/books/NBK92773/

55. White F (2015). Primary Health Care and Public Health: Foundations of Universal Health Systems. Medical Principles and Practice, 24(2): 103–116.

56. Who global report on traditional and complementary medicine 2019. (2019).

57. WHO. WHO Traditional Medicine Strategy 2002–2005. Geneva: 2002

58. Willcox M, Siegfried N, Johnson Q (2012). Capacity for clinical research on herbal medicines in Africa. Journal of Alternative and Complementary Medicine, 18(6), 622–628. https://doi.org/10.1089/acm.2011.0963

59. World Health Organization (WHO) (2002). WHO Traditional Medicine Strategy 2002–2005. World Health Organisation Geneva, 1–74. http://www.wpro.who.int/health_technology/book_who_traditional_medicine_strategy_2002_2005.pdf

60. Yang L, Wen KS, Ruan X, Zhao YX, Wei F et al. (2018). Response of Plant Secondary Metabolites to Environmental Factors. Molecules (Basel, Switzerland), *23*(4), 762. https://doi.org/10.3390/molecules23040762

61. Yuan Y, Tang X, Jia Z, Li C, Ma J, Zhang J (2020). The Effects of Ecological Factors on the Main Medicinal Components of *Dendrobium officinale* under Different Cultivation Modes. Forests. 2020; 11(1):94. https://doi.org/10.3390/f11010094

62. Zhang J, Onakpoya IJ, Posadzki P, Eddouks M (2015). The safety of herbal medicine: from prejudice to evidence. Evidence-Based Complementary and Alternative Medicine: eCAM, *2015*, 316706. https://doi.org/10.1155/2015/316706

2

Research and Development Efforts in Traditional Medicine Towards Achieving Universal Health Coverage in the African Region

Julius Ossy Muganga Kasilo[1], Peter Bai James[2], Kofi Busia[3]

[1]*Traditional Medicine, Medicines Supply, Health Infrastructure, Equipment Maintenance including Health Technologies, Medicines and Traditional Medicine Unit, WHO/AFRO*
[2]*National Centre for Naturopathic Medicine, Faculty of Health, Southern Cross University Researcher, New South Wales, Australia*
[3]*Principal Postgraduate Supervisor, Faculty of Medicine, Lincoln University College, Malaysia*

2.1 Introduction

The use of medicinal plants as a fundamental component of the African traditional healthcare system is perhaps the oldest and the most assorted of all therapeutic systems. Since Africa is known to be the cradle of civilization, ATM is the oldest form of traditional health practice in the world. In many parts of rural Africa, traditional medicine practitioners (TMPs) prescribing medicinal plants or phytomedicines are the most easily accessible and affordable health resource available to individuals, families, and local communities and sometimes the only therapy that subsists (Mahomoodally, 2013).

In addition, phytomedicines are also good sources of new products and innovation. For example, from 1981 to 2014, there were 1,562 newly approved therapeutic agents for all diseases worldwide. Of the 175 small molecules approved for the treatment of cancer from 1940 to 2014, 131 (75%) were naturally derived. For other disease conditions, the influence of natural products is quite marked, with, as expected, anti-infectives being dependent on natural products and their derivatives (Newman and Cragg, 2016).

For example, for the medicinal plant *Catharanthus roseus* L. (Madagascar periwinkle), its bitter and astringent leaves are used as emetic and its roots as purgative, vermicidal, depurative, haemostatic, and analgesic. In Mauritius, the juice of the leaves is used for indigestion and dyspepsia. In West Indies and Nigeria, the plant is used for diabetes, whereas in most parts of Africa, the leaves are used for menorrhagia and rheumatism. In Malaysia, *Catharanthus roseus* is used to treat diabetes, hypertension, insomnia, and cancer. In America, the gargle of the plant is used to ease sore throats, chest ailments, and laryngitis, whereas in India, the juice of the leaves is used topically to treat bee or wasp sting. In the Philippines, the decoctions of mature leaves and young leaves are used respectively for diabetes and stomach cramps, while the root decoction is used for intestinal parasitism. Furthermore, the infusion of the leaves is used for treating menorrhagia. In addition, the crude leaf extracts and root of *Catharanthus roseus* has anticancer properties, whereas the roots are used for dysentery.

Laportea ovalifolia is used in Cameroon to treat diabetes (Tsabang et al., 2015). Other plants such as *Momordica charantia, Morinda lucida, Moringa oleracea, Pausinystalia yohimbe,* and *Prunus africana* are commonly used. Such plants with established clinical effects have been actively studied in laboratories of research institutions and universities. Meanwhile, almost 90% of African countries which suffer from biodiversity depletion, climate change, and emergence and re-emergence of human diseases depend on foreign pharmaceutical industries and laboratories and could, therefore, benefit from these phytomedicines.

Although there is very little documented information in other parts of Africa in the precolonial era, as early as 3000 BC, the ancient Egyptians had a great deal of confidence in the efficacy of plants for treating many diseases. Africans depended mainly on TM for their healthcare needs. In Egypt, the famous Ebers Papyrus, written in 1550 BC, lists 842 prescriptions that include 328 different ingredients. Among them are plant species that grow in Egypt or in other North African countries, such as *Artemisia absinthium* (Compositae), *Acacia* spp (Mimosaceae), *Balanites aegyptiaca* (Balanitaceae), *Bryonia* spp (Cucurbitaceae), *Hyoscyamus muticus* (Solanaceae), *Myrtus communis* (Myrtaceae), *Onopordon* spp (Asteraceae), and *Ziziphus* spp (Rhamnaceae) (WHO, 2014; Batanouny, 2002). Dioscorides, born in the first century AD in Anazarba, a town in northern Cilicia (southeastern Asia Minor), in his *materia medica*, gave the names of many plants from Egypt, such as *Acacia nilotica* (Leguminosae), the Egyptian thorn, and Cyrenaica (e.g. *Dorema ammoniacum*, Umbelliferae).

Istifan ibn Basil (Stephen, son of Basil), translated the *materia medica* to Arabic, which was revised by the Syrian physician Hunayn ibn Ishaq in Baghdad in the ninth century. Herbalists in the Islamic world, which then extended from Central Asia to Andalusia, wrote many books and treatises on medicinal plants. Among them was the *Continens* (al Hawi fi'Tibb), which consists of twenty volumes on therapeutics. One of the Moslem scholars who was born and lived in North Africa was Ibn El Jazzar al Quairawani (died AD 1005). He wrote many books, including one on simple medicines. This book describes 272 medicines, many of plant origin, and has been translated to Greek, Latin, and Hebrew (Batanouny, 2002).

During the colonial era, African TM was deliberately rejected by the colonialists while conventional medical practices were forcibly imposed. The introduction of conventional medicine led to a systematic neglect of African TM, which adversely retarded its development. The neglect is partially attributed to the laws prohibiting the practice of TM by the colonialists, the assimilation of Western lifestyle by younger generations, and the lack of documentation. However, TM was still practiced especially in rural communities where it formed the main primary healthcare service for the people. Despite limited investment in the development of TM, many scientists compiled inventories of ethnobotanical, ethnomedical, and ethnoveterinary information on African plants. For example, in West Africa, Dalziel (1956) and Oliver (1959) made such contributions to the ethnobotany of the region. Similar work was done by other scientists, including Bally (1937) and Brenan and Greenway (1949) in East Africa.

These authors documented the uses of plants for medicine, food, fodder, timber, oils, and other aesthetic uses (Eyong, 2007). Another major contribution to the ethnobotany of the continent was *The Useful and Ornamental Plants in Zanzibar and Pemba* (Williams, 1949), which gives an ethnobotanical documentation of plants used for medicine, food, condiments, beverages, insecticides, perfumery, dyes, and rubbers, among others. Watt and Breyer-Brandwijk (1962) compiled an authoritative book with thousands of literature references on work done in Eastern and Southern Africa, covering most of the first half of the twentieth century. This book is a very valuable source of ethnopharmacological information on medicinal and poisonous plants in the Eastern and Southern African regions. In addition, TM usage for the prevention and treatment of diseases continued to be widespread across Africa. Among the reasons for the continued patronage of African TM include alignment with sociocultural, religious, and spiritual values, dissatisfaction with conventional medicines, as well as perceived low cost.

2.2 Political commitment and advocacy of African leaders and experts to promote research and development in Africa

With the advent of political independence, Africans rediscovered their sociocultural identity and realised that TM was an integral part of their heritage. Economic circumstances were also making imported medical equipment and medicines less accessible (Akerele, 1991). Furthermore, the sustained efforts of the WHO regarding the institutionalisation of TM were crucial to a policy shift by the countries. This advocacy effort was also demonstrated by African leaders at national, regional, continental, and global levels through their political commitment to promoting TM by the adoption of decisions, resolutions, and declarations that promoted the development of TM in the region.

At the regional level, the forty-ninth session of the WHO Regional Committee for Africa (RC49) by its resolution AFR/RC49/R5 on *the Essential Drugs in the WHO African Region: Situation and Trend Analysis* (WHO, 1999) requested WHO in 1999 to support member states in carrying out research on medicinal plants and promoting their use in healthcare delivery systems. The nineteenth session of the African Advisory Committee for Research and Development in 2000 recommended that the regional office should revitalise research on TM, particularly for common problems such as HIV/AIDS, tuberculosis, malaria, and childhood illnesses.

Furthermore, by resolution AFR/RC50/R9 on *Promoting the Role of Traditional Medicine in Health Systems: A Strategy for the African Region* (WHO, 2000), the Regional Committee urged member states in 2000 to *inter alia* (a) produce evidence on the safety, efficacy, and quality of TMs and undertake relevant research and (b) document medicines of proven safety, efficacy, and quality and facilitate the exchange and utilisation of this information by the countries. That resolution also requested the WHO to strengthen identified collaborating centres and other research institutions to carry out research, develop monographs of medicinal plants, and disseminate results on the safety and efficacy of TM products. In addition, the fifty-seventh Regional Committee for Africa in 2007 declared TM research as a priority. In 2008, the *Algiers Declaration on Research for Health* recognised the need to promote research in TM and strengthen health systems, considering the sociocultural and environmental situation of the people.

Moreover, by resolution AFR/RC63/R3 on *Enhancing the Role of TM in Health Systems: A Strategy for the African Region (2013–2023)* (WHO, 2013), the WHO Regional Committee for Africa urged member states in 2013 to *inter alia* (a) take concrete steps to assess the funding needs for TM research and development (R and D) and allocate adequate financial resources from national budgets while considering innovative funding sources and mechanisms and (b) invest in biomedical and operational research aimed at expanding the scope of the accepted best practices of TM in national health systems. The resolutions also requested WHO to (a) develop guidelines for the documentation of medicines of proven safety, efficacy, and quality and facilitate the exchange and utilisation of this information by the countries and (b) provide technical support for TM R and D to generate evidence and knowledge and promote the innovation and local production of TM products for priority diseases.

At the continental level, in 1964, the Organization of the African Unity (OAU) (now African Union [AU]) established the Scientific, Technical and Research Commission (STRC). This commission organised its first international conference on African TM in 1968 in Dakar, Senegal, where it was resolved that TM products dispensed by THPs should be evaluated to establish their claims. The priority areas identified in the screening of medicinal plants to provide scientific evidence for claims of efficacy included anticancer medicines and anthelmintic, antihypertensive, antimalarial, antimicrobial, antisickling, antiviral, and cardioactive agents.

Evidently, this decision was taken because very little research was being carried out on African medicinal plants then. Subsequently, the WHO, UNIDO and other institutions had undertaken many initiatives. The research that had been previously undertaken had only studied the general chemistry of the plant constituents. Moreover, the OAU/STRC had carried out ethnobotanical surveys in Cameroon, Ghana, Uganda, Swaziland, and Western Nigeria (Adjanohoun et al., 1993; Adjanohoun et al., 1996). The outcomes of these ethnobotanical surveys had been published and could be consulted in databases such as those developed by the OAU/STRC in Lagos, Nigeria, by the Natural products for Eastern and central Africa (NAPRECA) and the NARISTAN database on medicinal plants developed by Hoechst/ Naristan and used by the University of Cape Town (Sofowora, 1999).

The OAU/STRC has thus funded seventeen research centres all over Africa to stimulate research in this virgin area of proof of efficacy of medicinal plants in the region. These initiatives have greatly enhanced medicinal plant research (OAU/STRC, 1999) for the development of some phytomedicines, and the conduct of ethnobotanical surveys. Information on medicinal plant use has been obtained from herbalists, herb sellers, and indigenous people in Africa over the years (Baba et al., 1992; Wondergem, 1989). For example, under the umbrella of the *Agence de Co-operation Culturelle et Technique*, Paris, ethnobotanical surveys have been carried out jointly with international teams in Benin, Central African Republic, Comoros, Congo, Gabon, Madagascar, Mali, Mauritius, Niger, Seychelles, Tunisia, Togo, and Rwanda (Sofowora, 1996).

In 2001, the OAU heads of state and government declared at its summit in Abuja that research on TM should be made a priority. Furthermore, the OAU Summit held in Lusaka declared the period 2001–2010 as the *Decade for African TM* (OAU, 2001) and in 2003 adopted a plan of action for its implementation. During the same year, the African Summit of Heads of State and Government endorsed the WHO decision on the institution of the African TM Day for advocacy, celebrated on 31 August of every year. The AU Conference of African Ministers of Health held in Windhoek in 2011 discussed the End-of-Decade Review report on African TM and renewed the Decade from 2011 to 2020. One of the priority interventions of the first and second decades of African TM as well as the above mentioned WHO regional strategies, is R and D.

At the global level, in 2003, the Fifty-Sixth World Health Assembly adopted Resolution WHA56.31 on TM, which urged member states to provide adequate support for research on traditional remedies. That resolution requested WHO *inter alia* to (a) provide technical support for the development of methodology to monitor or ensure product quality, efficacy, and safety, the preparation of guidelines, and the promotion of exchange of information; (b) seek, together with WHO collaborating centres, evidence-based information on the quality, safety, efficacy, and cost-effectiveness of traditional therapies so as to provide guidance to member states on the definition of products to be included in national directives and proposals on traditional medicine policy as used in national health systems; (c) ensure the safety, efficacy, and quality of herbal medicines by determining national standards for, or issuing monographs on, herbal raw materials and TM formulas; and (d) collaborate with other organisations of the United Nations system and nongovernmental organisations in various areas related to TM, including research and protection of traditional medical knowledge.

In September 2008, the Beijing Declaration stated that TM should be further developed based on research and innovation in line with the 'Global Strategy and Plan of Action on Public Health, Innovation and Intellectual Property' adopted at the Sixty-First World Health Assembly in May 2008. In addition, in 2009, the Sixty-Second World Health Assembly adopted Resolution WHA62.13, which urged member states to further develop TM based on research and innovation, giving due consideration to the specific actions related to TM in the implementation of the global strategy and plan of action on public health, innovation, and intellectual property. The resolution requested WHO *inter alia* to (a) continue providing technical guidance to support countries in ensuring the safety, efficacy, and quality of TM, considering the participation of people and communities and taking into account their traditions and customs; (b) strengthen cooperation with WHO collaborating centres, research institutions, and nongovernmental organisations to share evidence-based information, taking into account the traditions and customs of indigenous peoples and communities; and (c) support training programmes for national capacity building in the field of TM.

It is interesting that the WHO and countries' advocacy efforts culminated in the establishment of research institutes to study the quality, safety, and efficacy of medicinal and aromatic plants. The oldest research institute was established in 1958 in Madagascar (Malgache Institute of Applied Research [IMRA]), which was followed by the National Chemotherapeutics Research Laboratory established in Uganda in 1963. Additional national research institutes have been gradually established, with emphasis on TM research. These include the National Institute of Phytotherapy and TM, whose name changed to the National Institute for Public Health Research in Mali (1968); the National Centre for the Applied Pharmaceutical Research (Madagascar, 1971); the Institute of Traditional Medicine of the Muhimbili University College of Health Sciences, now Institute of Traditional Medicine of the Muhimbili University of Allied Health Sciences (MUHAS) (United Republic of Tanzania, 1974); the National Research Institute (Blair) (Zimbabwe, 1974); and the Centre for Scientific Research into Plant Medicine (Ghana, 1975) (WHO, 2005), among others. In this article, we report the regional status of the R and D of African TM using information from surveys conducted in forty-seven countries in sub-Saharan Africa (SSA) carried out in 1999/2000 as baseline and in 2002, 2005, 2008, 2010, 2012, 2016, 2018, and 2020, as well as initial efforts for homegrown solutions with traditional medicine against COVID-19 during 2020–2022.

The surveys focused on indicators related to R and D to document scientific evidence on the safety, efficacy, and quality of African TMs used for the treatment of five priority diseases (HIV/AIDS, malaria, diabetes, hypertension, and sickle-cell disease). The development of monographs and herbal pharmacopoeias, patents of TM products (this aspect is reported in this book in the article on 'Leveraging on local production of TM to accelerate attainment of UHC'), established or strengthened research networks and research partnerships, increased government/public research funding for TM research, and collaboration between THPs and CHPs were also documented. Relevant information from different WHO fora that reviewed progress in the R and D of African TM in SSA was factored into the questionnaire (WHO, 2004, 2005). The baseline information collected from 2000 to 2018 was considered postcolonial, and this was complemented by the literature review to obtain information on the precolonial (prior to 1880) and colonial (1880–1960) period, which has been included in the background.

2.3 WHO support to member states on R and D and facilitation of regular exchange of experiences

In implementing the TM resolutions mentioned above, member states have been undertaking research to generate relevant scientific evidence on TMs used for the treatment of priority diseases. Evaluation of quality, safety, and efficacy based on research is needed to improve approaches to assessment of phytomedicines, a situation made difficult to remedy in light of historically inadequate public and private funding to address this growing concern.

Research and development (R and D) focuses on the efficacy, safety, quality, durability, strength, and reliability of the medical products. Regulatory systems' strengthening of phytomedicines has enabled national medicines regulatory authorities to provide an independent review of both the clinical and the preclinical data relating to phytomedicines and other medical products. Registration of new medical products and medical devices is a mandatory requirement before these become accessible to the public. R and D of African TM leading to standardised and quality-assured phytomedicines impacts positively on the three objectives of UHC. Access to phytomedicines in poor communities and villages promotes equitable service delivery.

However, validation of the efficacy and safety of TM products requires special research methodologies (WHO, 2000). WHO has provided support in this area, especially through the Special Programme for Research and Training in Tropical Diseases (TDR). To undertake credible R and D activities on TM, the WHO Regional Office for Africa prepared appropriate research methodologies and guidelines regarding various aspects of the scientific and clinical validation of phytomedicines used for the treatment of the five priority diseases. Member states agreed to these research methodologies at regional fora organised by WHO in Antananarivo, Madagascar, in November 2000 (WHO, 2000) and adopted in Harare, Zimbabwe, in November 2001 (WHO, 2004). The research tools were subsequently adopted by the

WHO Regional Expert Committee on TM for adaptation by member states to their specific situations.

Subsequently, some research institutes field-tested the guidelines, and the WHO Regional Office for Africa, in collaboration with the TDR, convened regional and intercountry fora in Johannesburg, South Africa, in 2002 (WHO, 2002) and 2004 (WHO, 2004) and in Nairobi, Kenya, in 2005 (WHO, 2005). In addition, the WHO Regional Office for Africa convened regional workshops on R and D and intellectual property rights in conjunction with the third meeting of the WHO Regional Expert Committee on TM in Johannesburg in 2004 (WHO, 2004) and in Harare, Zimbabwe, in 2012 (WHO, 2012) and 2014 (WHO, 2015). The objectives of these regional fora were to promote collaboration, exchange of country experiences on the use of the protocols and harmonisation of research methodologies, dissemination of best practices, and review of the progress attained in the R and D of standardised phytomedicines used for the treatment of the five selected priority diseases. Evidently, the adoption, adaptation by member states to their specific situations, and use of the WHO guidelines led to the WHO publication of the guidelines and protocols for the five diseases, which were combined into one guideline (WHO, 2004).

The research efforts, supported by the WHO, led to the development of potential antimalarial agents from indigenous plants undertaken at the Kenya Medical Research Institute, Nairobi; the Centre for Scientific Research into Plant Medicine, Mampong-Akwapem, Ghana; and the National Institute for Pharmaceutical R&D, Abuja, Nigeria. In 2002, the TM Programme of the Department of Essential Medicines at the WHO headquarters and the regional office co-organised a workshop in Accra, Ghana, to review the preliminary research results of the antimalarial agents studied by the three institutions. The reports revealed the potential therapeutic relevance of the antimalarial agents extracted from three different medicinal plants. It was therefore recommended that the institutions should continue with further R and D on these antimalarial agents.

2.4 Results on country progress on R and D to produce scientific evidence on the safety, efficacy and quality of phytomedicines used for the treatment of priority diseases

It should be noted that the earlier fora paved the way for the WHO African region to undertake the above mentioned surveys that would determine the country research status on African TMs for the management of the selected priority diseases. These surveys revealed an incremental number of countries conducting research to evaluate the safety, efficacy, and quality of African TMs used for the management of the selected priority diseases. By 2019, thirty-four research institutions from twenty-six countries reported their research efforts on phytomedicines against the five selected priority diseases using WHO guidelines as compared to eighteen in 2000. Their details are summarised below in Figure 1. Some countries used research results for policy change by issuing marketing authorisations for some standardised phytomedicines used for communicable and non-communicable diseases and included them in national essential medicines lists.

2.4.1 Clinical research on development of antimalarial medicines from African medicinal plants

By 2018, the above mentioned surveys revealed that twenty-two countries, *Benin, Burkina Faso, Cameroon, Chad, Comoros, Democratic Republic of the Congo, Ethiopia, Gabon, Ghana, Guinea, Kenya, Madagascar, Mali, Mauritius, Mozambique, Nigeria, Rwanda, South Africa, Uganda, the United Republic of Tanzania, Zambia,* and *Zimbabwe,* were conducting clinical evaluation of traditional herbal medicines used for the treatment of malaria applying the WHO protocols. The results of the phase I and II clinical trials in which the test subjects and control subjects were given herbal medicines and standard treatment with conventional medicines, respectively, showed that these herbal medicines cleared malaria parasites and increased haemoglobin within five to seven days, in many cases with little or no observable side effects. Some of the phytomedicines have gone through further clinical research and obtained marketing authorisations, whereas for others, further investigations are at different stages of progress in these countries. Information on the comparative effectiveness and safety profiles of the herbal medicines and comparator medicines will be useful here.

For example, research institutes in West Africa such as Benin, Burkina Faso, Ghana, Mali, and Nigeria have been evaluating the safety and efficacy of *Amaranthus graecizans L*, which has resulted in the development of a phytomedicine called Agbaya, used for the treatment of uncomplicated malaria in Benin (WHO, 2014). It is noteworthy that the R and D efforts in Burkina Faso has yielded two phytomedicines for the treatment of uncomplicated malaria, Saye and N'Dribala (Benoit-Vical et al., 2003), caused by *Plasmodium falciparum* and *Plasmodium berghei* parasites and *Plasmodium falciparum*, respectively. Saye is manufactured from three medicinal plants: the roots of *Cochlospermum planchonii, the leaves of Cassia alata, and the* leaves *Phyllanthus amarus*, whereas N'Dribala is from the roots *of Cochlospermum planchonii*. The two phytomedicines were registered in 2016 and included in the national essential medicines list by the national authorities of Burkina Faso.

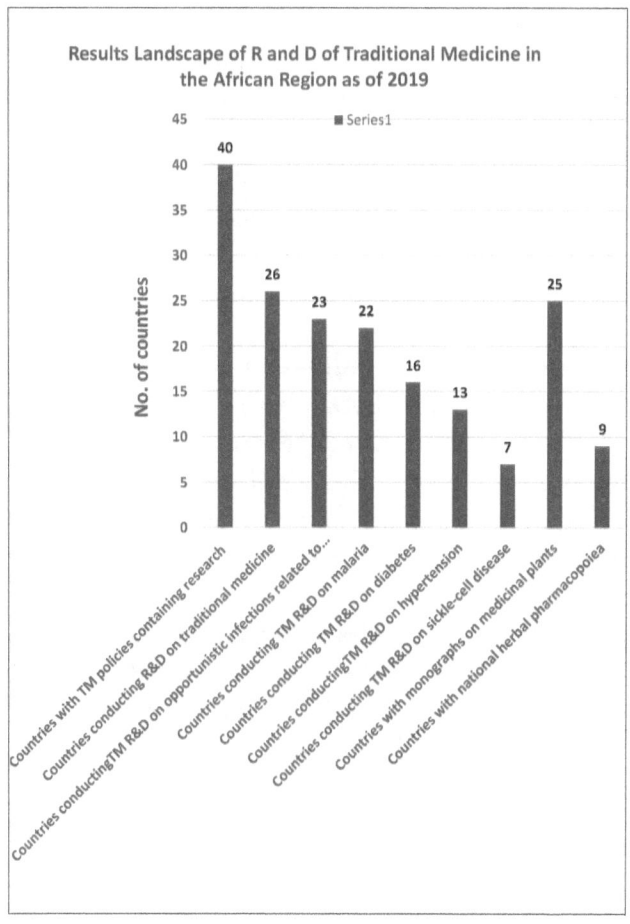

Similar results have been obtained in Ghana, where two antimalarials, Nasra and Nibima, have been evaluated by the Centre for Scientific Research into Plant Medicine (CSRPM), now Centre for Plant Medicine Research (CPMR) at Mampong-Akwapem, Accra, in pilot clinical trials, demonstrating *in vitro* and *in vivo* anti-plasmodial activities of both preparations (Addy, 2005; Oseni and Akwetey, 2012). A decoction prepared from the aqueous root extract of the plant *Cryptolepis sanguinolenta* has been used at the CPRM as an antipyretic and antimalarial for several years. The researchers at CPRM have shown that *Cryptolepis sanguinolenta* is not only antibacterial but also antimalarial, and it has been reported to contain active alkaloids such as cryptolepine and its derivatives. Patients clinically diagnosed with uncomplicated malaria (defined as parasitaemia of >1000/µl on thick blood film, together with clinical signs and symptoms) were followed up for forty-five days. Participants received a treatment of the herbal medicine Nibima at varied doses of 60–100 ml three times daily for seven days, after which the thick blood film for parasitaemia was repeated. Each participant was then followed up for recrudescence with thick blood film and full blood count (FBC) examinations on days 14, 28, and 45 (Thomford et al., 2014).

Cryptolepis sanguinolenta (Lindl.) is a plant well known for its prophylactic antimalarial potential, with the product sold under the trade name Nibima. The study seeks to establish the secondary benefit of using the product as an antimalarial, which was classified as the delay in recrudescence: absence of parasitaemia and cardinal signs of malaria such as fever and anaemia over a forty-five-day period. Ten participants were involved in the study, with a mean age of 32.70 (±17.45), comprising seven males and three females. Parasitaemia for all the participants were in the range of 103–104/µl, mean WBC (9.10 ± 0.20) and Hb (10.08 ± 1.29 g/dl). The repeat of the thick blood film on day 7 showed a total clearance of all malaria parasites, WBC (9.27 ± 2.16) and Hb (11.27 ± 0.27 g/dl). Monitoring during the follow-up period (days 14, 28, and 42) also indicated the absence of parasitaemia. Mean WBC and Hb was 4.64 ± 0.24 and 11.49 ± 1.44 g/dl, respectively, on day 42. The prevention of possible malaria recrudescence by the herbal product Nibima means that apart from the product being used primarily for the treatment of malaria, it may also hold some prophylactic potential that may need further investigation.

The work of Keita et al. (1996) on antimalarial preparations used in TM in Mali also indicates that a good number of herbal remedies can be proven scientifically to be effective. Seven herbal preparations for the treatment of various conditions, including malaria, have been included in the national essential drugs list of Mali. In addition, in the case study of Mali, a formulation containing three plants, Malarial, was subjected to three clinical trials for safety and efficacy in Baguineda (Wilcox et al., 2012). The second study was a randomised controlled clinical trial comparing chloroquine to Malarial that involved fifty-three patients, of whom seventeen were randomised to chloroquine and thirty-six to Malarial. The results indicated cure rates of 59% and 75% for chloroquine and Malarial, respectively. Although Malarial was better tolerated than chloroquine and fever clearance was similar in both groups, parasite clearance was better in the chloroquine group.

The third study on Malarial was an observational cohort study on patients with uncomplicated malaria, which involved thirty patients aged five years or above, with a temperature of >37.5°C and a parasitemia of >3000/mcl *P. falciparum*. There was no control group in this observational study. The Malarial used in this study contained an increased amount of *Acmella oleracea*, from 4% to 6%, because it was perceived that the 4% of *Acmella oleracea* present in the formulation used in the second study was insufficient for a truly effective schizonticidal activity. The parasitemia decreased, and the symptoms improved. Parasitaemia at day 7 remained higher in patients aged eight to nineteen than in older patients. This suggests that patient immunity was playing a role in clearing the parasites. The clinical findings showed that Malarial had significant antimalarial activity, resulting in its inclusion in the national essential medicines list as well as the Malian National Formulary alongside conventional medicines.

In Nigeria, the National Institute for Pharmaceutical Research and Development (NIPRD) developed a standardised antimalarial phytomedicine called NIPRD-AM-1 (Ameh et al., 2010). The pilot clinical trial on NIPRD 92/001/1-1, which involved three hospitals, was a comparative study using chloroquine and Fansidar as comparators and was carried out at the university teaching hospital and two general hospitals in Abuja. Chloroquine was used as the control drug at the specialist hospital and Fansidar at the two general hospitals. The clinical data indicated that NIPRD-AM-1 was superior to both Fansidar and chloroquine as regards parasite clearance and resolution of clinical symptoms (Gamaniel, 2009).

In Eastern and Southern Africa, the Kenya Medical Research Institute (KEMRI) also undertook a randomised controlled pilot clinical study on an antimalarial preparation which reduced parasites and fever and increased haemoglobin levels in five days (Orwa et al., 2012).

Sixteen patients were recruited for clinical investigation, with nine cases in the treatment group on neem tree leaf extract (*Azadirachta indica*) and seven in the control group on Fansidar. In the treatment group, seven cases (77.8%) showed complete cure (parasite clearance by day 7), and two cases (22.2%) showed partial cure (substantial reduction in parasitaemia by day 7). In the control group, six cases (85.7%) showed complete cure, with one case lost to follow-up. Parasite clearance was observed in most of the patients, without any observable adverse effects. Haematological parameters were normal, as were vital signs (WHO, 2002). On its part, the work of IMRA, Antananarivo, Madagascar, on malaria resulted in the discovery of a new medicine, Masy Totaquina, from *Cinchona ledgemia*, which is effective against a chloroquine-resistant strain of *P. falciparum* (Ranaivoravo, 2003). It is also used in flu and rheumatism.

2.4.2 Clinical research on the development of medicines used for opportunistic infections related to HIV/AIDS from African medicinal plants

The research and development of traditional medicines used for the management of opportunistic infections related to HIV/AIDS is a complex process, which involves many challenges. This process includes preclinical and clinical trials as well as an industrial evaluation, with the eventual marketing of the medicine meeting the established quality, safety, and therapeutic efficacy standards. Although numerous African plants have already been through preclinical evaluation/clinical trials with encouraging results, to date, there have been insufficient comparative clinical trials. Nevertheless, there have been encouraging developments in some countries with the objective of finding potential candidate medicines against HIV/AIDS. Based on these WHO-sponsored evaluation surveys, the following twenty-five countries indicated that they were conducting clinical evaluation of TM products used for the treatment of opportunistic infections related to HIV/AIDS: *Benin, Burkina Faso, Cameroon, Chad, Comoros, Cote d'Ivoire, Democratic Republic of the Congo, Ethiopia, Gabon, Ghana, Guinea, Kenya, Madagascar, Mali, Mauritius,*

Mozambique, Nigeria, Rwanda, Senegal, South Africa, Togo, Uganda, United Republic of Tanzania, Zambia, and *Zimbabwe.*

In all these countries, except Burkina Faso, Democratic Republic of the Congo, Nigeria, Senegal, South Africa, Uganda, and Zimbabwe which have conducted clinical trials, only clinical observational studies have been carried out because of ethical considerations and financial constraints. In Burkina Faso, two medicines have been developed, with successfully completed phase II clinical trials, especially through the identification of medicinal plants that can have a negative reaction with antiretroviral therapies (Nikiema et al., 2010). WHO supported ethnomedical studies conducted by institutions in Burkina Faso and Zimbabwe where, apart from baseline CD4/CD8 and viral load values measured at the inception of the study and reassessed every three months, liver and kidney function tests were undertaken using specific protocols (Simpore, 2001). Results showed that some herbal preparations reduce viral load (WHO, 2001). In addition, improvements had been noted in the quality of life and clinical conditions of patients treated with locally produced medicines. Blood tests to monitor the level of immunity (CD4 and CD8 counts) of patients, all of whom were treated exclusively with TM products, showed a marked increase in blood cell counts. In some countries such as Burkina Faso, a weight gain of up to 20 kilograms was noted in some patients within four months of treatment.

Similarly, in the United Republic of Tanzania, a herbal preparation prescribed by the Munufu Traditional Medicine Research Clinic in Dar es Salaam called Muhanse M4 has been in use since 1987 for the management of people living with HIV/AIDS. The rationale of providing treatment and care for HIV/AIDS patients is to prolong the lives of people living with the condition, ensure that the young generation is free of it, and to overcome stigma. Muhanse M4 has a long history in herbal medicine for boosting the immunity of people in various parts of Tanzania, particularly in the Mufindi and Kilombero districts. Muhanse M4 is just a standard infusion or weak decoction of ready-prepared powder from the whole plant or its parts. In spite of the current availability of modern antiretrovirals (ARVs), this indigenous medicine is in continual use among local communities because of its affordability (Mhame, 2004).

In Zimbabwe, members of the Zimbabwe National Traditional Healers Association (ZINATHA) have, since 1993, collaborated with some research and training institutions to evaluate the impact of phytomedicines in persons infected with HIV. The quality of life and HIV progression, including socio-demographic characteristics, were assessed. The researchers concluded that phytotherapy helps improve the quality of life of HIV-1 infected patients, but its pharmacological basis is still unknown (Sebit et al., 2001). Similarly, research into African TM for the management of opportunistic infections related to HIV/AIDS was carried out at the University of Zimbabwe, leading to the local production of the herbal medicine Gundamiti, with up to 97% reduction of the viral load in patients with HIV/AIDS. Gundamiti also had an appreciable effect on the decrease in HIV/AIDS-related opportunistic infections without affecting kidney and liver function (Kasilo, 2003).

A local pharmaceutical firm in South Africa has standardised into tablets a herbal preparation, Sunderlandiaâ, that is used as a tonic for diseases associated with significant loss of body mass. The water extract of the plant is used traditionally for the treatment of anxiety, asthma, bronchitis, cancer, depression, diabetes, gonorrhoea, influenza, and reflux oesophagitis. Another product being sold in South Africa for case management of PLWA is *Hypoxis rooperii* (also known as the African potato), as well as a variety of herbal preparations under the name of health food supplements, such as Sutherlandia and spirulina. These are traditionally used to treat chronic viral and bacterial diseases, along with different forms of *aloe vera*. THPs have been using it to treat cancer of the bladder and prostate and sexually transmitted diseases (STDs). Studies done on the plant have shown that it contains sterols and sterolins, which are essential dietary fats or lipids (Mills et al., 2005). The plant has helped many people to recover quicker from chronic and other diseases. It is a partly poisonous root, but with the right preparation and dosage, it is an approved immune booster to assist the body's natural defence system.

The University of Stellenbosch (South Africa), for instance, has conducted extensive research on this medicinal plant and has developed easy-to-take tablets. There are indications that although the African potato (*Hypoxis rooperii)* is not a sufficient treatment on its own, it could be extremely helpful when combined with other forms of treatment. Studies in the same university on HIV care revealed that the plant had shown the ability to increase the quality of life of patients, increase CD4 counts (the amount of white blood cells in the body), increase the weight of patients, and decrease the amount of HIV in the body (De Klerk, 2004). A collaborative study between companies such as PhytoNova and the Nelson Mandela Hospital in Durban, South Africa, resulted in the development of immune boosters.

2.4.3 Clinical research on the development of medicines used for treating fungal infections associated with HIV/AIDS from African medicinal plants

In Tanzania, Uganda, and Zimbabwe, some local medicinal plants are claimed to be used effectively for treating fungal infections associated with HIV/AIDS. In Uganda, CHPs and THPs have been working together since 1992 under the auspices of an NGO, THETA (traditional and conventional health practitioners working together against AIDS), to conduct research on medicines potentially useful for combating HIV-related illnesses. One such herbal preparation was used for the treatment of herpes zoster on which THETA has conducted some controlled clinical trials. Comparing subjects treated with herbal medicines with controls using acyclovir, the conventional treatment for herpes zoster, both groups were found to experience similar rates of resolution of herpes zoster. The investigators reported that the TM group had less superinfection and showed less keloid formation compared to those patients on acyclovir. Furthermore, pain due to herpes zoster reduced significantly faster in the group on herbal medicine compared with the group on acyclovir. It was concluded that herbal treatment is an important local and affordable means of managing herpes zoster in HIV/AIDS-infected patients in Uganda (WHO, 2001).

2.4.4 Clinical research on development of phytomedicines used for noncommunicable diseases from African medicinal plants: diabetes, hypertension and sickle-cell disease

Research institutions in sixteen African countries, *Benin, Burundi, Cameroon, Comoros, Democratic Republic of Congo, Ethiopia, Ghana, Guinea, Madagascar, Mali, Mauritius, Nigeria, Rwanda, South Africa, the United Republic of Tanzania*, and *Zimbabwe*, are involved in the clinical evaluation of the safety, efficacy, and quality of phytomedicines used for the management of diabetes. For example, Ghana has been working on medicinal plants that have indicated benefit to patients by saving glycogenolysis and reducing vascular complications caused by low-density lipoproteins. Their work is at different stages, with encouraging results. The Institute of Health Sciences Research (*Institut de Récherche en Sciences de la Santé*) in Burkina Faso and the *Institut Malagache des Recherches Appliquées* in Madagascar have developed phytomedicines for the treatment of diabetes mellitus (Ratsimamanga and Jambolana, 2019).

Thirteen countries (*Benin, Cameroon, Cote d'Ivoire, Ethiopia, Ghana, Guinea, Kenya, Madagascar, Mali, Mauritius, Nigeria, Rwanda,* and *South Africa*) in the African region are evaluating the safety, efficacy, and quality of phytomedicines used for the treatment of hypertension. For example, researchers in Nigeria reported that R and D work on hypertension has resulted in two medicinal products, which are undergoing clinical trials, and another product for which phytochemical and toxicological research have been carried out (WHO, 2006).

In addition, Mali is conducting research on the potential antihypertensive agents of the extracts. Research organisations created for the development of TM, supported in this area by the WHO, are involved in activities dealing with the main illnesses such as malaria, HIV/AIDS, sickle cell disease, diabetes, and hypertension. The treatment of hypertension with conventional medicines is too expensive, leading many patients to use TMs. It is essential that these should be safe, effective, and of proven quality. Different plant organs used by traditional practitioners have been the subject of diuretic phytochemical and antihypertensive studies at the Department of TM in Bamako and the Scientific and Health Research Institute in Ouagadougou. These include *Cymbopogon giganteus, Gynandropsis gynandra, Portulaca oleracea, Jatropha gossypiifolia*, and a traditional practitioners' recipe (Diallo et al., 2019).

For example, the infusion of *Portulaca oleracea* at a dose of 37.5 mg/kg, with a urinary excretion rate of 163.10%, produced substantial diuretic activity in Mali. A rise in tension provoked by a 75µg/kg dose of adrenalin was inhibited by the aqueous maceration of *Jatropha gossipiifolia* 94.64% to a dose of 20mg/kg. But very few phytomedicines from African TM have obtained approval for commercialisation, with the exception of Guinex-HTA, produced in Guinea.

Research and development of antisickling medicines is a priority in Africa, with the highest incidence of sickle cell disease. The antisickling medicine FACA (a combination of *Fagara xanthoxyloides* and *Calotropis procera*) was developed in Burkina Faso, starting from a traditional medicine and with the support of the WHO Regional Office for Africa. Its antisickling, antiinflammatory, antipyretic, and muscle relaxant properties as well as toxicity were tested. The plants that make up FACA act synergistically against the principal symptoms of the sickle cell disease. Administered under clinical conditions, FACA is well tolerated and significantly reduces the frequency of crises (Nikiema et al., 2010). After gaining approval for commercialisation, FACA is now being produced industrially, and it has been included in the national essential medicines list.

R and D work by some Nigerian researchers from Awolowo University, Ile-Ife, in Nigeria on the use of Fagara in managing sickle cell disease is another illustrative example of research on noncommunicable diseases. The active ingredients were characterised after demonstrating *in vitro* the antisickling activity of the roots of *Zanthoxylum zanthoxyloides*. Various R and D efforts on this plant were published, including standardisation work that led to its inclusion as Fagara in the African Pharmacopoeia. Similarly, R and D works on *Cryptolepis sanguinolenta* by a research team at the CPMR in Mampong-Akwapim, Ghana, led to the isolation of the active ingredients. Pharmacological and toxicological studies as well as standardisation work on this plant led to its inclusion in the Ghana herbal pharmacopoeia.

The NIPRD in Abuja, Nigeria, has developed a multiple-herb product that has been approved in Nigeria for the treatment of sickle cell disease (Wambebe, 2001). It is a natural product formulated from four medicinal plants, three of which are essentially food plants (*Piper guineense* [Piperaceae], *Pterocarpus osun* [Fabaceae], *Eugenia caryophyllata/Syzigium aromaticum* [Lauraceae], and *Sorghum bicolor* [Poaceae]), which grow wild or are cultivated in West Africa. This plant-based medicine was used by THPs in Nigeria and has recently been the subject of *in vitro* studies and clinical trials (Swami et al., 2009). *In vitro* studies showed that the product protected laboratory animal models of sickle cell disease from death when exposed to low oxygen tension. Furthermore, clinical data showed that about 70% of patients were protected from crises, while the frequency and severity of crises in the remaining volunteers were dramatically reduced. The product (Niprisan) has been patented in the USA and 40 other countries worldwide. Other plant formulations used for the disease are being evaluated by the Esoma Herbal Research Institute and Neimeth in Nigeria.

The research by the team at NIPRD, led to the development of a herbal medicinal product called Niprisan, which is meant for the management of sickle cell disease. That marked an important shift in matters concerning patent as compared with the publication of research findings on the continent. Data accumulated on Niprisan up to the multicentred phase 3 clinical trials were used to strengthen the patent protections already filed in some forty countries by the UNDP on this plant product. Similar studies on sickle cell disease are being conducted in Benin, Burkina Faso, Togo, and Nigeria.

2.5 Establishment or strengthening of research networks and partnerships and databases

Despite the benefits of research networks, including interdisciplinary expertise, funding opportunities, and technology transfer, less than 45% of the forty-seven countries in the African region have developed research networks in TM R and D. Furthermore, less than 30% have developed research databases that provide a knowledge management advantage. Consequently, there is very little information about the progress made in different aspects of TM. Research databases offer the advantage of knowledge management such that research type and results are harnessed in a systematic manner and made available to enhance knowledge evaluation, utilisation, and management. It is important that countries pursue networking with institutions in other countries that have the requisite capacity and facilities for successful research and development activities. Research networks provide the benefit of collaboration between different disciplines and institutions within and outside the countries in a collaborative effort around specific research. In addition, research networks offer the advantage of interdisciplinary expertise, infrastructure required to conduct complex R and D studies, shared learning, shared resources, better funding opportunities, and technology transfer.

Examples of countries whose research institutions have either established or strengthened research networks and partnerships include *Benin*, which has partnerships with the West African Health Organization (WAHO), PROMETRA Benin, and research institutes on naturopathy. Burkina Faso has links and collaborates with the Association Nassara, Association for the Burkina Society of Ethnopharmacology and Ethnobotany, and WAHO. Côte d'Ivoire has links and collaborates with WAHO; UNICEF; Alliance Côte d'Ivoire; FHI 360; and CPMR in Mampong-Akwapem, Ghana; Kwame Nkrumah University of Science and Technology in Kumasi, Ghana; universities and national research institutes; and THPs within the country. Ghana established a memorandum of understanding with research institutions and Mozambique with the Chinese government, private sector, and NGOs. Madagascar has partnerships with manufacturers, the ministry of public health, and Salama central medical store, whereas Mali has partnerships with the Geneva Foundation for Medical Education and Research. Nigeria partners with the private sector in local production. The United Republic of Tanzania has several links with scientists in the Institute of Traditional Medicine of the University of Allied and Health Sciences in Dar-es-Salaam and National Institute of Medical Research. The country has entered into an agreement with the Chinese government to assist in capacity building of research scientists.

2.6 Increased government/public research funding for TM research

About 30% of the respondent countries provided research funding for TM research. Table 1 on government/public research funding for TM research shows that only Benin, Ethiopia, Mali, Madagascar, Mozambique, Nigeria, and the United Republic of Tanzania provided government allocations for TM research.

Table 1: Government/public research funding for traditional medicine research in USD

Years	Benin	Ethiopia	Madagascar	Mali	Mozambique	United Republic of Tanzania
2010–2011	260 000	250 000		101 888	125 000	Not available
2012–2013	360 000	250 000		10 000	125 000	Not available
2014–2015	500 000	320 000		48 296	125 000	180 132
2016–2017	400 000	440 000		50 000	125 000	349 772
2017–2021			83 928.58			
Total	1 520 000	1 260 000		110 000	1 000 000	529 904

2.7 Collaboration between THPs and CHPs in research and development

Enhanced referral systems of patients from THPs to CHPs and vice versa, resulting from better communication and collaboration of these two types of practitioners, also contribute to improving access to essential health services by rural dwellers where most of the THPs practice, and this is in line with UHC. Within the context of PHC, the practitioners of the two systems can blend together in a beneficial harmony using the best features of each system and compensating for certain weaknesses in each. The THP–CHP collaboration involves, among others, training of THPs in PHC and early signs for referrals of their patients. Similarly, the CHPs become conversant with African TM pathophysiology of diseases and diagnostic techniques and can then engage in case studies with CHPs. Consequently, such a practice promotes quality of health services by both THPs and CHPs, which aligns with one of the objectives of UHC.

In Kenya, the nongovernmental organisation (NGO) Women Fighting AIDS in Kenya (WOFAK), in collaboration with THPs, have used TM products for managing their HIV/AIDS patients since 1994 (UNAIDS, 2002). This NGO has identified useful herbal therapies for herpes zoster, diarrhoea, malaria, skin rash, cough, fever, and joint pains (Management Sciences for Health, 2012). Since the early 1990s, the Tanga AIDS Working Group (TAWG), an NGO based in Tanga in the United Republic of Tanzania, has been promoting collaboration between THPs and conventional health practitioners in the management of HIV/AIDS patients. As of 2013, TAWG had treated over 6,000 AIDS patients with opportunistic infections using herbal medicines.

TAWG has reported promising results with improved quality of life of their patients weight gain, and control of most symptoms associated with the disease. Four medicinal plants have been identified through this collaborative work, and standardisation work is in progress with the Institute of Traditional Medicine of the University of Dar-es-Salaam and the National Institute of Medical Research of the Ministry of Health and Social Welfare in the United Republic of Tanzania (Kayombo et al., 2007). In November 2013, the National Institute for Medicine Research (NIMR) and Ministry of Health in the United Republic of Tanzania announced the positive development of an herb, Tashack, which is a promising candidate for the management (Itala, 2013) of opportunistic infections related to HIV/AIDS.

In a similar manner, the NGO Traditional and Modern Health Practitioners Together against AIDS (THETA) in Uganda supported clinical observations with 194 volunteers in 1992 and had encouraging results. Treatment with standardised herbal medicines on ten patients showed the same effect as the THP formulations. Preparations used by THPs for treating people living with HIV/AIDS have led to a herbal product that improves the CD4 count of AIDS patients. The NGO also identified TM remedies that relieve people living with HIV/AIDS of diarrhoea and vomiting. Based on these experiences, institutions of higher learning and research in, for example, Kenya, Uganda, and the United Republic of Tanzania, are currently at various stages of development of preparations for HIV/AIDS. It is believed that sooner or later, promising candidates with antiviral activity may be achieved from these ongoing research activities (UNAIDS, 2002).

In Nigeria, NIPRD has reported two of the many herbal preparations THPs claim to be effective for the management of HIV/AIDS: Dopravil and Conavil. Dopravil was used for the management of 23 volunteers who participated in a retrospective study. These findings were confirmed by a prospective pilot clinical trial conducted by NIPRD using only thirteen patients. In December 2000, NIPRD officials reported that the mandatory clinical trial of Dopravil would be completed within two years, subject to the availability of funds. At that time, over 10,000 PLWA had been treated with Dopravil out of compassion. The phase II clinical study on Conavil began in December 2000, whose extracts are also used for both prophylactic and symptomatic treatment of upper respiratory tract infections, lower urinary tract infections, and acute diarrhoea. The preliminary data obtained showed that all the volunteers benefited from the treatment with no reported side effects, while all HIV/AIDS-related symptoms noted prior to therapy were diminished. Baseline CD4/CD8 and viral load values were estimated at the inception of the study in December 2000 and reassessed every three months, starting from March 2001.

Furthermore, PROMETRA International's Experimental Centre for Traditional Medicine (CEMETRA) in Fatick, Senegal, which consists of 450 member associations of THPs of Sine, known as Malango, officially recognised by the government of Senegal, THPs collaborate with Western-trained medical doctors. PROMETRA as well as THPs and a research institute in the United States of America have been collaborating in conducting clinical research. CEMETRA shelters the healthcare units run by THPs and a modern office managed by a qualified physician. An important characteristic of CEMETRA is that only THPs are authorised to treat patients within the centre. The interview gives an insight into the background of the patient before the physical examination that follows. The doctor never administers a modern treatment. He only makes a diagnosis before, during, and after the treatment given by the THP. Clinical examination is on the various functioning of the body. Preliminary check up includes assessment of blood pressure, pulse, respiration, temperature, and weight. The laboratory equipment at CEMETRA is credible and is supervised by a certified technician. Various clinical examinations help the doctor to confirm his diagnosis. Once the modern diagnosis is established, the doctor refers the patient to the qualified THP. This later ignores the diagnosis made by the physician, and he has his own method of diagnosis.

It should be noted that the treatment used by the THP is broad based and varied. The treatment could take the form of a bath, ingestion of herbs, sacrifice, or ritual. The patient returns to the modern structure for a repeat clinical and laboratory evaluation. This allows for a post-treatment analysis, which determines if there was worsening, no change, or improvement in the patient's disease. Through these collaborations, practitioners of the two systems of medicine have enhanced mutual confidence and trust, as well as referral of patients from THPs to modern doctors and vice versa.

In addition, the collaboration between THPs and conventional health practitioners in Senegal have enabled PROMETRA International to produce and patent a pharmaceutical product based on medicinal plants, METRAFAIDS, used for the treatment of opportunistic infections related to HIV/AIDS. In a clinical observational study, three cohorts of 62 HIV positive individuals (18 men and 44 women) were treated with METRAFAIDS. More than half of the patient population (54%) had a viral load decrease of greater than 66% without any adverse reactions throughout the study (Amzat and Abdullahi, 2008; Addy, 2005). This collaboration helped to reduce health workers' scepticism and strengthened mutual appreciation, understanding, and respect between practitioners of the two health systems of medicine.

2.8 Development of monographs and herbal pharmacoppoieas

Monographs, national and regional pharmacopoeias define quality specifications and standards for herbal materials and some herbal preparations, such as essential oils and powdered herbal materials. Use and inclusion of herbal materials in such pharmacopoeias are based on local availability of these products. Availability is dependent on the original medicinal plants, which have ecologically and environmentally specific habitats. Therefore, even if the same pharmacopoeial monograph name is given to a herbal material, its listing in one pharmacopoeia may refer to a different original medicinal plant and/or processing method from that defined in another (WHO, 2004).

It is interesting to note that except for Egypt, Mozambique and Sudan (currently divided into Sudan under the East Mediterranean Regional Office [EMRO] and South Sudan under the WHO-AFRO Regional Office) where herbal pharmacopoeias (and monographs in Mozambique) are legally binding, all the national herbal pharmacopoeias are not legally binding. Since pharmacopoeias set quality control and quality assurance standards for African medicinal plants and herbal medicines, they should serve as legal resources. This could be readily achieved if a national or regional authority expressly introduces a national pharmacopoeia or any part of it into appropriate legislation. These pharmacopoeias are vital reference tools for all individuals and organizations involved in herbal and pharmaceutical research, development, manufacture, quality control, and analysis.

For example, in the Economic Community of Central African States in Cameroon, national monographs on herbal medicines are used but are not legally binding. A monograph on plants is used to treat some priority conditions: diabetes, diarrhoea, sickle cell disease, hypertension, malaria, and tuberculosis. Other monographs on herbal medicines used are the *WHO Monographs of Selected Medicinal Plants* volume 3 (2007) and volume 4 (2009). The Democratic Republic of Congo published its first edition of the herbal pharmacopoeias in 2009 (Congo Ministry of Health, 2009), whereas Equatorial Guinea has national monographs contained in the *Recetario plantas medicinales de Equatorial Guinea* (1996), which contains 18 monographs (WHO, 2005), whereas national monographs in Guinea exist in the *Plantes médecinales guinéennes* (1997) (WHO, 2005).

By 2018, twenty-five countries (Benin, Burkina Faso, Cameroon, Chad, Congo [Republic], Côte d'Ivoire, Democratic Republic of Congo, Equatorial Guinea, Ethiopia, Gambia, Ghana, Guinea, Madagascar, Mali, Mauritius, Mozambique, Niger, Nigeria, Rwanda, Senegal, Seychelles, Sierra Leone, South Africa, Togo, and Uganda) had developed monographs on medicinal and aromatic plants (MAPs), an increase of twenty-three countries compared to the baseline survey of 1999/2000. For example, in the East Africa community, Uganda had developed the national pharmacopoeia entitled *A Contribution of the TM Pharmacopoeia of Uganda* (1993) (WHO, 2005), while other countries in EAC use *WHO Monographs on Selected Medicinal Plants* volume 4 (2007). In some countries of the Economic Community of West Africa States (ECOWAS) such as Benin, national monographs on herbal medicines are used but are not legally binding. Examples include two 2009 monographs on plants used in TM in Benin for the prevention and treatment of malaria and sexually transmitted diseases (STDs) and HIV/AIDS. The *Ghana Herbal Pharmacopoeia* was published in 1992 (GHP, 1992) and 2007 (GHP, 2007); and the *Nigerian Herbal Pharmacopoeia* was published in 2008 with WHO support.

Between 2008 and 2020, the Regional Office for Africa had supported the development of two volumes of the *West African Health Organization (WAHO) Herbal Pharmacopoeia* published in 2013 and 2020, respectively (WAHO, 2013), for the ECOWAS member states, spearheaded by WAHO. The *WAHO Herbal Pharmacopoeia* contains commonly used and geographically distributed medicinal plants in West Africa for treating the priority diseases selected by the WHO Regional Office (together with tuberculosis and hepatitis and for which some information exist in the African Pharmacopoeia) (WHO, 1999/2002). An African Pharmacopoeia containing a dictionary and multilingual monographs of potential African medicinal plants of eight West African countries has been published in two volumes (Eklu-Natey and Balet, 2012).

Similarly, in some countries of the Southern African Development Community (SADC) such as Mozambique, the national monographs are contained in the series:

- *Medicinal plants of traditional use in Mozambique (Plantas medicinais e seu uso tradicional em Mocambique)* comprising five monographs (five volumes) published in 1983, 1984, 1990, 1991, and 2001;
- *Ethnobotanical research on medicinal plants in the province of Manica and Zambezia (Pesquisa ethnobotanica sobre plantas medicinas na provineia de Manica e Zambezia)* comprising two monographs issued in 2001 and 2004; and
- *Medicinal plants used in the treatment of diseases (Plantas medicinais utilizadas no tratamento de doenmentais)*, a single monograph published in 2009.

In 2010, the South African Pharmacopoeia Monograph Project was underway, with sixty-three pharmacopoeias and monographs listed, while Madagascar uses *Towards a Malagasy Pharmacopoeia (Vers une pharmacopee Malagasy)*, part 1 published in 2008.

2.9 Efforts for research and development of traditional medicines against COVID-19

During the COVID-19 pandemic, several countries in the World Health Organization (WHO) African region proposed several traditional medicine-based remedies for the treatment of COVID-19. WHO, in collaboration with the Centre for Diseases Prevention and Control (Africa CDC) and the African Union Commission for Social Affairs (AUC), jointly established the Regional Expert Advisory Committee on Traditional Medicine for COVID-19 Response (REACT) in July 2020. The REACT was established to (a) accelerate the pace of conducting research and development (R and D) and applying standards of clinical trials of TM-based therapeutics for COVID-19 through pooling expertise for multicentre studies; (b) provide independent scientific advice; (c) support countries to generate scientific evidence on the safety, efficacy, and quality of TMs; and (d) create the conditions for consolidating data across the African continent.

To guide countries in assessing the safety and efficacy of the TM-based therapeutics, the REACT developed generic protocols for conducting randomised, multicentre clinical trials as well as clinical observational studies of herbal medicines. At least ten countries, namely, Burkina Faso, Democratic Republic of Congo (DRC), Equatorial Guinea, Ghana, Guinea, Madagascar, Nigeria, South Africa, Togo, and Uganda, adapted these protocols to their specific situations to conduct phase 2 and 3 clinical trials of TM-based therapeutics against COVID-19. Thereafter, WHO organised field missions to these countries to monitor the clinical trials being conducted. The mission selected some few herbal medicine-based therapeutics whose clinical trials need to be finalised to determine the safety, efficacy, and quality for fast tracking for local manufacture. However, funding was a big challenge mentioned by all countries, which could impede research and development of these products.

2.10 Challenges

Despite these achievements, there are challenges that need to be overcome, including the following:

- Paucity of updated comprehensive compilation of promising medicinal plants from the African continent
- Lack of technological and scientific capabilities to participate in commercial collaborations and opportunities created by the Biodiversity Convention
- Lack of financial resources and infrastructure to reduce the burden of diseases
- Limited expertise and poor access to the necessary technological infrastructure for research discovery on phytomedicines
- Absence of strategies to carry out botanical bioprospecting for producing new medicines and other products
- Insufficient documentation of African TM
- Difficulty in verifying the training of any of the present THPs
- Inadequate information and limited data on research, especially scientific evidence relating to safety and efficacy
- Difficulty elucidating the chemical complexity of plant-based formulations due to limited relevant equipment and expertise
- Lack of appropriate research methodologies for TM research without undermining its main ethos
- Inadequate allocation of domestic financial resources for research and development of traditional medicines

2.11 Conclusion

The prospects for enhancing African phytomedines require addressing the policy and legal framework, with special attention paid to capacity building for R and D, funding, intellectual property rights, and advocacy. Many countries in WHO's African region lack the capacity in terms of human resources and research facilities, to carry out focused and credible R and D on medicinal plants. Concerted efforts must be made at the levels of the member states, WHO, and development partners to address this. The capacities of centres of excellence (COE) can be strengthened to enable researchers to undertake relevant research activities.

Governments should allocate at least 2% of national budgets of health ministries to R and D agencies to implement the call to action issued at the Global Ministerial Forum on Research for Health held in Bamako, Mali, in November 2008 (http://www.who.int/rpc/news/BamakocalltoactionFinalNov24.pdf, 2019). In addition, member states should pursue innovative financing mechanisms for health research and link evidence to policy-making. WHO has provided technical support to member states and is committed to continue accordingly. The international development partners which have supported R and D for TM are highly commended. It is anticipated that such support would continue. In Africa, although patentable results have been obtained, many of these have not yet been exploited commercially. There is a need for a mechanism to induce commercialisation of novel findings, with appropriate benefit arrangements and processes possibly through OAPI and ARIPO.

African countries should utilise the *sui generis* system mechanism to protect TMs, genetic resources, and traditional medical knowledge. The OAU (now AU) Model Law on IPR and the WHO Regional Office for Africa's policy guidance (WHO, 2016) and *sui generis* legislative framework (WHO, 2016) for the protection of indigenous knowledge in African TM can assist member states to develop their national IPR regulations for the protection of their genetic resources. It is noteworthy that the above-mentioned IPR documents are complementary to the framework of the Global Strategy and Plan of Action on Public Health, Innovation and Intellectual Property, which charts the way forward for the R and D and protection of IPRs, adopted by the World Health Assembly (WHO, 2008).

There is the need for advocacy to sensitise policymakers, NGOs, biomedical scientists, THPs, and the communities, among others, on the importance of TM in the healthcare delivery system. Member states are urged to facilitate the institutionalisation of African TM in their healthcare delivery system. African countries should embark on advocacy for community orientation, dissemination of appropriate information, and promotion of positive attitudes and practices. In particular, it is necessary to involve all stakeholders in Africa so as to create awareness for the cultural relevance of African TM.

Subsequently, the programme for the development and promotion of African TM should be adequately publicised. Information on medicinal plants and African TM should be a core component of the curricula for health and allied health sciences students. Member states should endeavour to use success stories on the R and D of TMs in their advocacy programmes. Furthermore, the potential socioeconomic benefits of investing in the R and D of African TM and the comparative advantage of African bioresources, which are also renewable, should be articulated and shared widely among the stakeholders. To effectively undertake R and D of African TMs, a multidisciplinary approach is the only option to obtain new products. The collaboration should include internal, regional, and international colleagues as well as the private sector.

References

1. Addy ME (2005). Western Africa Network of Natural Products Research Scientists (WANNPRES), first scientific meeting august 15 -20, 2004. Accra, Ghana: a report. African Journal of Traditional, Complementary and Alternative Medicine, 2(2): 177–205. Accessed on 23 October 2019.
2. Adjanohoun E et al. (1993). Traditional Medicine and Pharmacopoeia: Contribution to Ethnobotanical and Floristic Studies of Uganda. OAU/STRC, Lagos.
3. Adjanohoun E. et al., (1996). Traditional Medicine and Pharmacopoeia: Contribution to Ethnobotanical and Floristic Studies of Cameroon. OAU/STRC, Lagos.
4. Akerele O (1991). Registration and Utilization of Herbal Remedies in some countries of East, Central and Southern Africa. pp. 3–8 In Proceedings of International Conference on Traditional Medicinal Plants, K.E. Mshigeni et al., (eds.) Arusha, United Republic of

Tanzania, February 18–23, 1990. Ministry of Health, United Republic of Tanzania.

5. Ameh S, Obodozie O, Gamaniel S, Abubakar M, Garba M (2010). Physicochemical variables and real time stability of the herbal substance of NIPRD-AM1®- an antimalarial developed from the root of *Nauclea latifolia* S.M. (Rubiaceae). International Journal of Phytomedicine 2: 332–340. Available at http://www.arjournals.org/ijop.html, accessed 25 August 2019. ISSN: 0975-0185.

6. Amzat J, Abdullahi AA (2008). Roles of Traditional Healers in the Fight Against HIV/AIDS. Journal of EthnoMedicine, 2(2): 153–159.

7. Baba S, Akerele O, Kawaguchi Y (Eds.) (1992). Natural Resources and Human Health. Tokyo: Elsevier.

8. Bally PRO (1937). Native Medicinal and Poisonous Plants of East Africa. Kew Bulletin 1, 10–26.

9. Batanouny KH (2002). Traditional medicine in North Africa: History, Importance, Status and Needs. P. 81–90. In: 2001-2010: OAU Decade for African Traditional medicine. Proceedings of the 15th Meeting of the Inter-African Experts Committee on African Traditional Medicine and Medicinal Plants, Adeniji, K.O. (Ed.), Arusha, United Republic of Tanzania, 15–17 January, 2002. OAU/STRC, Lagos.

10. Benoit-Vical F. et al. (2003). N'Dribala (Cochlospermum planchonii) versus chloroquine for treatment of uncomplicated Plasmodium falciparum malaria. Journal of Ethnopharmacology, 89(1):111–114.

11. Brenan JMP and Greenway PJ (1949). Check Lists of the Forest Trees and Shrubs of Tanganyika.

12. Busia K (2007) ed. Ghana Herbal Pharmacopoeia, 2nd Edition.

13. Dalziel JM (1956). The Useful Plants of West Tropical Africa. The Crown Agents for the Colonies, London.

14. David J, Newman DJ, Cragg GM (2016). Natural Products as Sources of New Drugs from 1981 to 2014. Journal of Natural Products; 793: 629–661, Available at: https:// doi. org/10.1021/acs.jnatprod.5b01055, accessed, 15 September 2019.

15. De Klerk P (2004). A potato, a traditional medicinal plant used to treat chronic viral and bacterial diseases and some forms of cancer,

is now recognized by physicians to boost the immune system of HIV-infected people. In: Hassan O. Kaya. Promotion of public health care using African indigenous knowledge systems and implications for IPRs: Experiences from Southern and Eastern Africa. 2007 African Technology Policy Studies Network (ATPS), Special Paper Series No. 30; ISBN: 9966-916-28-8.

16. Diallo D, Haidara M, Tall C, and Kasilo OMJ (2010). Recherche sur la médecine traditionnelle africaine: hypertension. Special Issue No. 14, Decade of African Traditional Medicin 2001–2010. Available at: http://apps.who.int/medicinedocs/documents/s21374en/s21374en.pdf, Accessed, 2 November 2019.

17. Eklu-Natey RD, Balet A (2012). Dictionnaire et monographies multilingues du potentiel médicinal des plantes africaines, Afrique de l'Ouest. Volume 1 et 2, Dictionnaire. Edition d'en bas, Lausanne - Traditions et Medecine T&M, Genève.

18. Ekpere JA (2000). The OAU Model Law: The Protection of the Rights of Local Communities, Farmers, and Breeders, and for the Regulation of Access to Biological Resources- An Explanatory Booklet. OAU/STRC, Lagos. pp. 100

19. Eyong CT (2007). Indigenous Knowledge and Sustainable Development in Africa: Case Study on Central Africa. In: Indigenous Knowledge Systems and Sustainable Development: Relevance for Africa, Boon EK and Hens L (eds), Kamla-Raj Enterprises, India, 121.

20. Gamaniel K (2009). A comparative randomized clinic trial of NIPRD AM1 against a chloroquine and sulphadoxine/pyrimethamine combination in symptomatic but uncomplicated malaria. In: Abstracts of the world congress on medicinal and aromatic plants, Cape Town, November 2008. African. Journal of Traditional, Complementary and Alternative Medicine, 6, 299–493. ISSN 0189-6016©2009.

21. Government of Ghana (1992). The First Edition of the Ghanaian Herbal Pharmacopoeia, 1992.

22. Itala M (2013). Tanzania: Tashack - a Promising Cure for HIV/AIDS. Tanzania Daily News (Dar es Salaam). Available at http://allafrica.com/stories/201311100191.html, accessed 15 May 2019.

23. Kasilo OMJ (2003). Enhancing traditional medicine research and development in the African region. Traditional medicine: Our Culture Our Future. African Health Monitor, 1: 5–18.

24. Kayombo EJ, Uiso FC, Mbwambo ZH, Mahunnah RL, Moshi MJ, Mgonda YH (2007). Experience of initiating collaboration of traditional healers in managing HIV and AIDS in Tanzania. Journal of Ethnobiology and Ethnomedicine, 3: 6. Published online 2007 January 26. doi: 10.1186/1746-4269-3-6.

25. Keita A (1996). Evaluation in vitro and in vivo of a traditional antimalarial: "Malarial 5". Journal of Alternative and Complementary Medicine, 2(3), 420.

26. Mahomoodally MF (2013). Traditional Medicines in Africa: An Appraisal of Ten Potent African Medicinal Plants. Evidence-Based Complementary and Alternative Medicine, Article ID 617459, 14 pages. Available at: http://dx.doi.org/10.1155/2013/617459

27. Management Sciences for Health (2012). Traditional and complementary medicine policy. In: Policy and regulatory policy. 5, 1–17.

28. Mhame PP (2004). The Management of HIV/ AIDS-Related Conditions Using A Traditional Herbal Preparation-Muhanse M4 In Tanzania: A Case Study In: Dar-Es-Salaam, Tanzania, National Institute For Medical Research. In: Hassan O. Kaya. Promotion of public health care using African indigenous knowledge systems and implications for IPRs: Experiences from Southern and Eastern Africa. 2007 African Technology Policy Studies Network (ATPS), Special Paper Series No. 30; ISBN: 9966-916-28-8.

29. Mills E, Cooper C, Seely D, Kanfer I (2005). African herbal medicines in the treatment of HIV/AIDS and Sunderlandia. An overview of evidence and pharmacology. Nutrition Journal, 4: 19.

30. Ministry of Health (2009). The First Edition of the Herbal Pharmacopoeia and Tradition. Kinshasa. Democratic Republic of Congo.

31. Nigerian Federal Ministry of Health (2008). Nigerian Herbal Pharmacopoeia. First Edition. In collaboration with the World Health Organization. ISBN: 978-978-088-020-0. Nigeria.

32. Nikiema JB, Ouattara B, Sembde, Djierro K, Compaore M, Guissou IP and Kasilo OMJ (2010). Promotion de la médecine

traditionelle du Burkina Faso : essai de développement d'un médicament antidrépanocytaire, le FACA. Traditional medicine: Our Culture Our Future. African Health Monitor 2003; 1: 52–57.

33. Nikiema JB, Simporé J, Sia D, Djierro K, Guissou IP and Kasilo OMJ (2010). L'introduction de plantes médicinales dans le traitement de l'infection à VIH: une approche réussie au Burkina Faso. Traditional medicine: Our Culture Our Future. African Health Monitor 2003; 1: 47–51.

34. OAU/STRC (1999). Proceedings of the Symposium on African Medicinal and Indigenous Food Plants and the Role of Traditional Medicine in Health Care. Includes the Report and Recommendations of the 14th Session of the Inter-African Committee of Experts on African Medicinal Plants and Traditional Medicine. OAU/STRC/University of Swaziland, Kwaluseni, Swaziland, 4–6 October 1999.

35. OAU/STRC (1985) African Pharmacopoeia. Volume 1, OAU/STRC, Lagos.

36. Wondergem P, Senah KA, Glover EK (1989). Herbal Drugs in Primary Health Care. Ghana: An Assessment of the Relevance of Herbal Drugs in PHC and Some Suggestions for Strengthening PHC. Amsterdam, Netherlands: Royal Tropical Institute.

37. OAU (2001) Decision on the Declaration of the period 2001–2002 as the OAU Decade for African Traditional Medicine. Doc. CM/2227 (LXXIV) Add.1

38. Oliver B (1959). Medicinal Plants in Nigeria. College of Arts, Science and Tech., Ibadan.

39. Orwa JA, Ngeny L, Mwikwabe NM, Ondicho J, and Jondiko IJO (2012). Antimalarial and safety evaluation of extracts from Toddalia asiatica (L) Lam. (Rutaceae). Journal of Ethnopharmacology, 11/2012. Available at: http://www.researchgate.net/institution/Kenya_Medical_Research_Institute/department/Centre_for_Traditional_Medicine_and_Drug_Research_CTMDR, accessed 12 January 2016).

40. Oseni LA, Akwetey GM (2012). An in-vivo evaluation of antiplasmodial activity of aqueous and ethanolic leaf extracts of *Azadirachta indica* in *Plasmodium berghei* infected balb/c mice.

International Journal of Pharmaceutical Sciences and Research (2012), Vol. 3, Issue 05, 2012.

41. Ranaivoravo J (2003). Laboratory and Clinical Evaluation of Traditional Medicines for Care and Treatment of Malaria: 33Experience of the Malagasy Applied Research Institute. African Health Monitor, World Health Organization Regional Office for Africa January - June 2003, 33 pp.

42. Ratsimamanga SU (2019). Eugenia Jambolana: Madagascar, Malagasy Institute of Applied Research, Antananarivo. http://ssc. undp.org/uploads/media/Eugenia_Jambolana Madagascar.pdf, accessed 1 October 2019.

43. Sebit MB, Chandiwana SK, Latif As, Gomo E et al (2001). Quality of life evaluation in-patients with HIV-1 infection: the impact of traditional medicine in Zimbabwe. Central African Journal of Medicine, 46(8): 208–213.

44. Simpore J (2001). Medicaments issus de la medecine traditionelle et traitement du VIH/SIDA au centre Saint Camille, Burkina Faso, presented at the Regional Meeting on Integration of traditional medicine in health systems, Zimbabwe, 26–29 November 2001.

45. Sofowora A (1999) Medicinal plants research in Africa: Prospects and problems, pp12–28. In: Proceedings of the Symposium on African Medicinal and Indigenous Food Plants and the Role of Traditional Medicine in Health Care. (Includes the Report and Recommendations of the 14th Session of the Inter-African Committee of Experts on African Medicinal Plants and Traditional Medicine). University of Swaziland, Kwaluseni, Swaziland, 4–6 October 1999.

46. Swami N, Prema T, Qingli W, Faith C (2009). Nicosan: Phytomedicinal Treatment for Sickle Cell Disease, In: African Natural Plant Products: New Discoveries and Challenges in Chemistry and Quality. 15, 263–276.

47. Thomford KP, Edoh DA, Thomford AK, Appiah AA (2014). The Prophylactic Potential of Nibima, a Ghanaian Herbal Antimalarial Product. Conference: West African Network of Natural Products Research Scientists (WANNPRES) At: Ile-Ife, Nigeria. July 2014.

48. Traoré M, et al. (2008). *In Vitro* and *In Vivo* Antiplasmodial Activity of 'Saye', An Herbal Remedy Used in Burkina Faso Traditional Medicine. Phytotherapy Research, 22, 550–55.

49. Tsabang N, Kadjo S, Mballa RN, Yedjou CG, Nnanga N, et al. (2015). New Approach for the Development of Improved Traditional Medicine: Case of a Preparation of an Oral Hypoglycemic Medicine from Laportea ovalifolia (Schumach. & Thonn.) Chew. (Urticaceae). Journal of Molecular Pharmaceutics and Organic Process Research, 3(2): 125. doi: 10.4172/2329-9053.1000125.

50. UNICEF/UNDP/World Bank/WHO Special Programme for Research and Training in Tropical Diseases (2002). A preparatory regional meeting held in Johannesburg, South Africa, 27–28 June 2002. In: Handbook on Non-clinical Safety Testing.

51. UNAIDS (2002). Ancient Remedies, New Disease: Involving traditional healers in increasing access to AIDS care and prevention in East Africa. UNAIDS Case Study (June 2002), UNAIDS Best Practice Collection. UNAIDS/02.16E (Original version, June 2002), ISBN 92-9173-171-4.

52. UNAIDS (2002). Ancient Remedies, New Disease: Involving traditional healers in increasing access to AIDS care and prevention in East Africa. UNAIDS Case Study (June 2002), UNAIDS Best Practice Collection. UNAIDS/02.16E (Original version, June 2002), ISBN 92-9173-171-4.

53. Wambebe C, Khamofu H, Momoh JAF, Audu BS et al. (2001). Double-blind, placebo-controlled, randomized cross-over clinical trial of NIPRISAN in patients with sickle-cell disorder. Phytomedicine, 8: 252–261.

54. Watt JM and Breyer-Brandwijk (1962). The Medicinal and Poisonous Plants of Southern and Eastern Africa. E. & S. Livingstone Ltd., Edinburgh & London, 1457.

55. West African Health Organization (2013). West African Herbal pharmacopoeia for Economic Community of the West African States (ECOWAS), Bobo Dioulasso, ISBN: 978-9988-1-8015-7, KS Printkraft Ghana, Ltd.

56. WHO (2014). Final Report of the Regional Workshop on development of country work plans for implementation of the

regional strategy on enhancing the role of traditional medicine in health systems. Harare, Zimbabwe, 11–14 May 2014.WHO Regional Office for Africa, Brazzaville, Republic of Congo (Unpublished).

57. WHO Regional Office for Africa AFR./RC49/R5(1999). Essential Drugs in the WHO African Region: Situation and trend analysis. Forty-ninth session, Windhoek, Namibia, 30 August -3 September 1999.

58. WHO Regional Office for Africa, AFR./RC50/9 (2000). Promoting the Role of Traditional Medicine in Health Systems: A Strategy for the African Region. Fiftieth session, Ouagadougou, Burkina Faso, 28 August - 2 September 2000.

59. WHO (2013). Enhancing the Role TM in Health Systems: A Strategy for the African Region. Sixty-third session of the WHO Regional Committee for Africa. WHO Regional Office for Africa, Brazzaville.

60. WHO (2005). Final Report of the Regional Workshop on Pre-Clinical Safety Evaluation of Traditional Medicines, Kenya, 25–30 August 2005. WHO Regional Office for Africa, Brazzaville (unpublished).

61. WHO (2004). Final Report of Regional workshop on Research and Development of Traditional Medicine and Intellectual Property Rights (Document AFR/TRM/04.2), held in Johannesburg, South Africa, 25–27 November 2004. Brazzaville WHO Regional Office for Africa.

62. WHO (2004). Pre-Clinical Safety Testing of traditional medicines, Johannesburg, South Africa September (Unpublished): WHO Regional Office for Africa, 2004.

63. WHO (2005). Final Report of the Consultative Meeting on Pre-Clinical Safety Testing of traditional medicines, Nairobi, Kenya, 25–30 August, 2005 (unpublished): WHO Regional Office for Africa, Brazzaville.

64. WHO (2012). Final Report of the Regional Workshop on research and development of traditional medicines, Harare, Zimbabwe, 21–23 May 2012. Brazzaville: WHO Regional Office for Africa, 2012.

65. WHO (2000). General Guidelines for Methodologies on Research and Evaluation of Traditional Medicine (Document, WHO/EDM/TRM/2000.1). World Health Organization, Geneva.

66. WHO (2000). Summary report of the Regional workshop on evaluation of traditional medicine, Antananarivo, Madagascar, 20–24 November. World Health Organization, Regional Office for Africa, Harare, Temporary location (unpublished).

67. WHO (2004). Final Report. Regional meeting on integration of traditional medicine into national health systems: Strengthening collaboration between traditional and conventional health practitioners. Harare, Zimbabwe, 23–26 November 2001 (Document, AFR/TRM/04/06).

68. WHO (2004). Pre-Clinical Safety Testing of TMs, Johannesburg, South Africa September. WHO Regional Office for Africa, Brazzaville (unpublished).

69. World Health Organization (2005). Final Report of the Consultative Meeting on Pre-Clinical Safety Testing of TMs, Kenya, 25–30 August, WHO Regional Office for Africa, Brazzaville (unpublished).

70. WHO (2004). Final Report of Regional workshop on Research and Development of Traditional Medicine and Intellectual Property Rights held in Johannesburg, South Africa, 25–27 November. WHO Regional Office for Africa, Brazzaville (Document, AFR/TRM/04.2).

71. WHO (2012). Final Report of the Regional Workshop on research and development of traditional medicines. Harare, Zimbabwe, 21–23 May 2012. WHO Regional Office for Africa, Brazzaville, Republic of Congo.

72. WHO (2015). Final Report of the Regional Workshop on development of country work plans for implementation of the regional strategy on enhancing the role of traditional medicine in health systems. Harare, Zimbabwe, 11–14 May 2014.WHO Regional Office for Africa, Brazzaville, Republic of Congo (AFR/EDM/TRM/2015.1).

73. WHO (2004). Guidelines for Clinical study of traditional medicines in the African Region. World Health Organization, Regional Office for Africa, Brazzaville.

74. WHO (2014). Final Report. Regional Workshop on development of country work plans for implementation of the regional strategy on traditional medicine, 11–14 November 2014. WHO Regional Office for Africa, Brazzaville Republic of Congo.

75. WHO Regional Office for Africa (2001). In: Wambebe C. (2000) The role of research in ensuring safety, efficacy of traditional medicines. Final Report of the African on role of traditional medicine in health systems, held in Harare, Zimbabwe, 15–18 February 2000 (AFR/TRM/01.1).

76. WHO Regional Office for Africa (2001) In : Kabatesi D (2000) The role of traditional medicine in HIV/AIDS prevention and control. Final Report of the African on role of traditional medicine in health systems, held in Harare, Zimbabwe, 15–18 February 2000 (AFR/TRM/01.1).

77. WHO (2006). Situation analysis of local production of traditional medicines in Nigeria, (Unpublished report).

78. (WO2002062364) antiviral composition made from medicinal plants for combating HIV/AIDS. WIPO Patent scope. Available at: http://patentscope.wipo.int/search/fr/WO2002062364, accessed 13 May 2019.

79. WHO (2004). WHO guidelines on safety monitoring of herbal medicines in pharmacovigilance systems. World Health Organization, Geneva. Available at: http://apps.who.int/iris/bitstream/ handle/ 10665/43034/9241592214_eng.pdf

80. WHO (2005). National policy on traditional medicine and regulation of herbal medicines. Report of a WHO global survey. [Online] Available at: http://apps.who.int/medicinedocs/ en/d/ Js7916e/9.1.html#Js7916e.9.1, accessed 28 April 2017.

81. WHO (1999/2002). WHO Monographs on selected medicinal plants. Volume 1 & 2. World Health Organization, Geneva.

82. WHO (2016). WHO Regional policy guidance for protection of indigenous knowledge in African traditional medicine. WHO Regional Office for Africa, Brazzaville (Document AFRO/ HTI/ TRM/2016.3).

83. WHO (2016). Regional sui generis legislative framework for the protection of traditional medical knowledge and access to

biological resources. Regional Office for Africa, Brazzaville (Document AFRO/HTI /TRM/2016.4).

84. WHO (2008). WHO Global strategy and plan of action on public health, innovation and intellectual property, WHO Resolution WHA61.21, World Health Organization, Geneva.

85. Willcox M, Sanogo R, Diakite C, Giani S, Paulsen S, Diallo D (2012). Improved traditional medicines in Mali. Journal of Alternative and Complementary Medicine, 18(3): 212–220.

86. Williams RO (1949). The Useful and Ornamental Plants of Zanzibar and Pemba. St. Ann's Press, Great Britain. 497.

3

Bridging the Ecological Link with Nature with Phytomedicines

Charles Katy[1], Kofi Busia[2]

[1]*Freelance Traditional Medicine Expert, Dakar-Senegal*
[2]*Principal Postgraduate Supervisor, Faculty of Medicine, Lincoln University College, Malaysia*

3.1 Introduction

Today, the industrial concept of medicine has led to the standardisation of treatment models for all disease conditions. Undoubtedly, medicine is becoming increasingly mechanical and business oriented and moving away from the holistic concept of health, which entails a balance between body and mind to achieve optimal well-being. In a new policy paper published in the *Annals of Internal Medicine*, the American College of Physicians is reported to have said that 'profit motive in medicine may contribute to a bloated, complex, and fragmented health care system' (ACP, 2021).

Indeed, modern medicine is being driven by the quest for profit, with huge returns on investment, which is facilitated by health policies that usually benefit the pharmaceutical and vaccine industries at the expense of community health. The Goldman Sachs Investment Bank's report of April 10, 2018, 'Keeping people sick for more profit than curing them', is instructive. Richard J. Roberts, the 1993 Nobel Prize Laureate for Medicine, said it all, 'Chemical pills chronicise disease but do not cure' (Roberts, 2007).

The Commission to the Council and European Parliament noted in its report of February 2014 that 'in many countries, health is one of the most corrupt sectors' (COM, 2014). The reign of the pharmaceutical industry, which is usually more concerned with its financial health than with human health, makes it increasingly imperative to return to the fundamental principle of medicine, which according to Hippocrates is: 'First do no harm'. This must also be reinforced by his warning to all doctors: 'Nature has its own healing power! Nature is the physician of the sick'. Paracelsus added, 'The physician must not boast too much . . . He must know what nature wants and that nature is the first physician' (Paracelsus, 1520).

Though conventional medicine uses natural bioactive compounds to produce phytomedicines, the fact remains that the immunity of humans, is first and foremost, innate and natural and cannot be supplanted by that provided by synthetic laboratory medication. Phytomedicine, which proceeds from the scientific study of the medicinal properties of specific plants or plant extracts, invites the world to extricate itself from the trap in which the pharmaceutical industrial revolution confines it. In other words, when for health, it is a question of re-establishing the link with nature that has been broken in favour of the chemical laboratory. For, 'believing that the pharmaceutical industry fights against disease is like believing that arms manufacturers fight for peace'!

Through its scientific approach, phytomedicine accounts for the properties of plants and at the same time attests to the natural services provided by biodiversity, the mechanisms of which are part of an ecological engineering that establishes the balance of the relation among living beings in a universe where everything stands. Industrial chemistry must not kill these natural services with its fertilisers, pesticides, and herbicides, nor with its pollutants, which is the cause of many hormonal disorders, the burden of which is unprecedented in the history of wildlife and humans. By promoting green chemistry, phytomedicine is also a way of trusting nature more than technology.

A new paradigm towards medicine for the future must be built on a new approach to health that considers the relationship between the biological and the risks of industrial growth on the living. This calls for ecological engineering, which aims at understanding and preserving the mechanisms of the natural balance of the ecosystem, including relationships between living beings. The resilience of the ecosystem is due to the pollutants introduced by the petrochemical, pharmaceutical, agricultural, and food industries. In addition to these pollutants, our digital era is also exposed to electromagnetic waves. Electro-sensitivity is the cause of new diseases as bacteria and viruses use waves in their DNA and can amplify the damaging reception of vibrations on the human organism. Phytomedicine remains a plea for medicine to listen to the balance of the natural ecosystem for a better understanding of its ecological engineering. Health also partly depends on biodiversity structured by mechanisms to be preserved for the future of humans.

3.2 Physiological/phytobiological analogy of plants

The earth's ability to support life, including human life, has been shaped by life itself (National Academy of Science, 2017). Nature does not sleep. It is self-organised. With or without humans, plant species appear and disappear. The universe of the plant world takes us back to the origins of human life. Without trees, there would have been no soil, no water, no biodiversity, and indeed no life.

The question is have we wisely accorded the world of plants, which accounts for 82% of all life on our planet, the desired respect?

What have we learned by physiological analogy from the secrets revealed to humans by medicinal plants? Considering the stress of their ecosystems, which change through pollution, deforestation, or agriculture, how do plants react to their environment, of which humans are a part? The answer to these questions lies in the importance of phytobiology, which studies the mechanisms of the function of plant organs and tissues within their genetic pool. These mechanisms reveal a whole engineering of the relationship between ecology and plant medicine.

The evolutionary creativity inherent in nature (Sheldrake, 2001) is masterfully demonstrated by the communication network known as 'wood wide web'. Indeed, an immense underground network structures the circulation of all the molecules that govern the life of plants. While the sequencing of nuclear DNA provides knowledge on the evolution of human genome, phytobiology allows us to discover the secret life of trees through their language and intelligence. As social beings with emotions, pain, and anguish, trees maintain a community life of mutual aid that allows them to 'grow in a natural way, between different species, get aged and cultivate their offspring' (Dordel, 2016). Trees feed off each other so that the older ones take under their armpit the young shoots that connect to them.

Matter is structured energy, and plant protoplasm is no exception to this fact. By natural laws, networks of filaments formed by fungi in contact with the roots of trees order the distribution of carbon, nitrogen, phosphorus, hydrogen, and other derivatives of photosynthesis. The dynamics of such cellular self-organisation is based on the constant energy/matter associativity characteristic of dissipative structures (Prigogine, 1992).

The process of photosynthesis, by which humans are provided with oxygen to breathe, is emblematic of the vitality of our link with plants, which in turn absorb the carbon dioxide we exhale. Are humans not mere descendants of organisms that were able to adapt to oxygen? The reality of the dissipative structure of the plant kingdom requires humans to adapt to nature, which has the right answers to the issues concerning our survival as a species. Thus, phytomedicine takes a participatory approach by acting according to what nature offers and not according to what we want to do with it.

Deoxyribonucleic acid (DNA) is the genomic material in cells that contains the genetic information used in the development and functioning of all known living organisms. DNA, together with ribonucleic acid (RNA) and proteins, is one of the three major macromolecules that are essential for life. The order of genetic information is the basis of the ecological balance of nature and its flora. By analogy, the structure of the alkaloids present in plants is very close to that of the amino acid serotonin (Brelet, 2004), which carries information between nerve cells in the brain and throughout the body. In plants, this simple organic compound, alkaloid, harmonises the transport of genetic message from DNA to RNA, and also that of the bacterial reverse transcriptase from RNA to DNA (Beljanski, 2011). In the event of disease, an imbalance occurs and disrupts the order of genetic message transport.

Like nature, human organism is self-organised to eliminate, recycle, and repair all the cells' components. As demonstrated by Yoshinori Ohsumi, 2016 Nobel Prize Laureate for Medicine, the apoptosis mechanism allows a cell not only to trap and destroy viruses and bacteria, but also to self-digest and self-destruct to protect the organism.

Consequently, through its therapeutic virtues, the plant acts in harmony with the organism, allowing it to be the actor in the restoration of its own balance and its own epigenetic programming. The human body thus acquires the ability to heal itself. Molecular biologist Mirko Beljanski has even demonstrated that natural plant extracts have the property of inhibiting the synthesis of cancer-promoting DNA and not healthy DNA. This suggests that when plant extracts exclusively target cancer stem cells, they also have the property of accompanying the apoptosis mechanism so that the cells cannot develop into cancerous tumours.

3.3 Phytomedicine and Medicine of the Living

Through communication between microorganisms inherent in species, the components of living beings serve each other in the name of ecosystem balance. The more we understand the mechanisms of the relationships between living beings, the better we know ourselves as an integral part of our environment so as to extend our humanity to all of nature. This is why the destruction of forests and the resulting loss of biodiversity is suicidal. Today, in the face of this ecological peril, some plants have become metallophytes, managing to extract from the soil polluting elements introduced by the chemical industry.

By virtue of their magnetic properties, necessary for the development of their therapeutic properties, plants know how to give the best of themselves for the establishment of their immune system, their growth, and their longevity. Thus, phytoextraction, which enables plants to adapt to polluted environments, deserves to be studied to free pollutants from the soil and, by extension, to purify the human organism, which is subject to toxins emanating from economic activity that generates all sorts of endocrine disruptors.

If, according to Pasteur, 'a disease is determined by a cause, which is itself determined by a biological mechanism that can be corrected by an active principle' (Gigon, 2014), it should be noted that medicine deals with human beings, not machines. In other words, for effective healing, it is first and foremost paramount to take an interest in the person before tackling the disease because 'the microbe is nothing, the terrain is everything', as Claude Bernard and Pasteur so aptly observed later on, both quoting Pierre Valentin Marchesseau, founder of orthodox naturopathy (https://www.signesetsens.com/psycho-biographie-pierre-valentin-marchesseau-fondateur-naturopathie.html). Homeopathy makes us understand that we do not treat the disease in man, but man in disease. Obviously, man is linked to his environment, the products of which he is above. Man exists not only in his physical dimension but also in his mental and spiritual dimensions. This holistic approach to man has been well documented in the cultures of traditional societies, particularly in Africa among the Mande, Gur, Kru, and Akan groups (Konadu, 2008).

The Industrial Revolution, with its prodigious leap, has trapped humans by distancing us from Mother Nature, with the result that 'health care has become increasingly business-oriented with more for-profit entities and private equity investments' (ACP, 2021). Yet plants are our partners in the universe, and we speak the same language. By scientifically studying the medicinal properties of plants, phytomedicine re-establishes our link with plants and exposes the much broader spectrum of activities of herbal medicine preparations, all of which have pleiotropic effects. It is no coincidence that the best results of pharmaceutical companies come from the synthesis of natural products in the manufacture of their medicines. If this is the case, then it is important that we recognise the primacy of nature over the all-chemical laboratory. Also present in the identification, treatment, and prevention of plant diseases, phytomedicine provides information on their properties in organising their immune defence. Plant pathology is thus proving to be a first-rate field of research that provides us with natural therapies for human diseases.

3.4 Conclusion

Quantum physics has enabled us to recognise our limits and our ignorance of the unknown life forms of the world in which we live. Quantum physics confirms three postulates that traditional medicine already knew: Mankind is in the image of the protoplasm to which he is linked, we are all connected to each other, and our body is made up of energy (Brelet, 2004). In short, matter is structured energy, and humans are no exception to this reality. According to Einstein's famous theory of relativity ($E = MC^2$), energy can be transformed into matter and vice versa. Thus, energy acts on matter, matter produces energy, and the mind produces energy, which in turn acts on matter, which is nothing other than an energy whose vibration has been reduced to being perceptible by senses.

This is why rational phytomedicine is practised with respect for mankind and components of the living world, of which plants occupy a central place in the natural balance of the ecosystem. It is easy to understand the encyclopaedia of the botanical universe of the communities that live from their genetic resources when, in the semantics of each name of a plant, we find by devotion the linguistic seat of a conception of the environment with which we live in symbiosis.

The laboratory and its study model of the synthetic molecule cannot account for the biochemical complexity of plants. Through its heuristic approach, phytomedicine brings us closer to the complexity of plant for a good understanding of their biochemistry with multiple simultaneous and positive effects on living organisms, whose organs communicate with each other according to the laws that govern all species of the living matrix.

The beginning of the twenty-first century is worrying, with its epidemiological fracture, its health and food scandals, the imminent end of antibiotics, and the seeming exhaustion of the pharmaceutical industry. But we have good reason to be hopeful because of the immense pharmacy of nature and our ability to survive in the vibratory field of the living world, as well as the plants with which we share the same cosmic placenta and which link us to the earth through their roots.

It is hoped that phytomedicines will continue to be pivotal to the medicine of the future, which reconnects with the art of healing based on its approach centred on the person and his or her holistic human dimension. The genius of plants is such that their cellular physiology is beneficial to our holistic health, which is achieved by linking the body's physical, emotional, mental, and spiritual integrity.

References

1. American College of Physicians (2021). Profit motive in medicine may contribute to a broken health care system. https://www. acponline.org/acp-newsroom/acp-says-profit-motive-in-medicine- may-contribute-to-a-broken-health-care-system. Accessed on 18-05-2022
2. Beljanski M, Beljanski M (2011). La santé confisquée, Guy Trédaniel Available at : https://www.editions-tredaniel.com/la- sante-confisquee-p-417.html. Accessed on 24-11-2022
3. Brelet C (2004). Médecines du monde, Histoire et Pratiques des Médecines Traditionnelles, R. Laffont
4. Commission Européenne (2014). Rapport de la commission au conseil et au parlement européen, Bruxelles, le 3.2.2014, COM (2014) 44 final. https://eur-lex.europa.eu/legal-content/FR/TXT/ PDF/?uri=CELEX:52014DC0044&from=hr
5. Dordel J (2016). L'intelligence des arbres, documentaire. Available at https://www.allocine.fr/film/fichefilm_gen_cfilm=256847.html. Accessed on 24-11-2022.
6. Gigon F (2014). Les plantes au service de la civilisation. Sante Nature Innovation. Available at : https://www.santenature innovation.com/les-plantes-au-service-de-la-civilisation. Accessed 24-11-2022.
7. Konadu K (2008). Medicine and Anthropology in Twentieth Century Africa: Akan Medicine and Encounters with (Medical) Anthropology, African Studies Quaterly, 10.
8. Marchesseau PV (1922). Le fondateur de la naturopathie contemporaine, en France. Signes & Sens. https://www. signesetsens.com/psycho-biographie-pierre-valentin-marchesseau- fondateur-naturopathie.html. Accessed 24-11-2022.
9. Prigogine I (1992). Processus dissipatifs dans la théorie quantique Rapports de physique, 219: 93-108. DOI: 10.1016/0370-1573(92)90128-M.
10. Roberts Richard J (2007). Le médicament qui guérirait tout ne serait pas rentable, Entrevue du Journal La Vanguardia, 2007. http://sam-menerveovb.over-blog.com/2016/08/

chroniciser-les-maladies-est-plus-rentable-que-les-guerir.html. Accessed on 24-11-2022.

11. Sheldrake R (2001), L'âme de la nature, (Collections Spiritualites) (French Edition). Albin Michel.

12. Temkin O (2012). Volumen medicinae paramirum of Paracelsus. Available at: https://www.scribd.com/document/256865827/ Paracelsus-Volumen-Medicinae-Paramirum. Accessed on 24-11-2022.

4

Deliberate Environmentally Conservative Medicinal Plants Cultivation – Value Chain Articulation

Isaac Kingsley Amponsah

1Department of Pharmacognosy, Faculty of Pharmacy and Pharmaceutical Sciences, Kwame Nkrumah University of Science and Technology, Kumasi, Ghana

4.1 Introduction

Since ancient times, plant resources have been used by different cultures to serve diverse purposes, including the provision of food, shelter, medicine, cosmetics, magico-religious applications, and other cultural practices. Their importance to mankind has evolved over the years, contributing significantly to the livelihood of several families and national economies (Schippmann et al., 2002; Namdeo, 2018). Among these plant resources, medicinal plants (MPs) play a central role, not only as traditional medicines used in many cultures, but also as trade commodities and sources of novel molecules for the discovery of drugs to treat diseases (Namdeo, 2018).

Medicinal plants are plants that possess therapeutic properties or exert beneficial pharmacological effect in humans or animals. Although the use of MPs for the treatment of diseases predates recorded history, their application as isolated and characterised compounds for modern drug discovery and development started only in the nineteenth century (Veeresham, 2012). The first plant-derived drug available in commercial quantities for therapeutic use is morphine, which was isolated from the opium puppy plant *Papaver somniferum*. The analgesic aspirin was also discovered as the first semi-synthetic pure drug derived from the natural compound salicin, isolated from the plant *Salix alba*. Several other drugs have since been discovered from medicinal plants, accounting for about 25% of prescription drugs. For many people globally, herbal products are their preferred method of treatment (Wachtel-Galor and Benzie, 2011; Dias et al., 2012). Currently, the global trade value of medicinal and aromatic plants (MAPs) is estimated at US$800 million per year, and this is projected to grow at a rate of 15%–25% to about US$50 trillion by 2050 (UN Comtrade, 2018).

However, this growing global enthusiasm for medicinal plant resources has resulted in an alarming depletion of important medicinal plant species from their natural habitats. Overexploitation is a growing problem for many African medicinal species in areas where population growth, lack of access to Western medicine, poverty, and growing markets fuel unsustainable harvesting practices (Van Andel et al., 2012). Majority of these plants are collected from the wild and are, therefore, being wiped out at an alarming rate. Demand for medicinal plants from the wild has increased by 8%–15 % per year in Europe, North America, and Asia in recent times (Jamshidi-Kia et al., 2018). This is compounded by a vastly increasing human population and extensive destruction of plant-rich habitats due to anthropological activities. In Ghana, for example, the slash-and-burn system of traditional farming results in the destruction of plant species and ecosystems due to serious environmental degradation (Dorm-Adzobu, 1982). Other factors contributing to medicinal plant loss, especially in sub-Saharan Africa, include collection and gathering of fuelwood, charcoal production, commercial timbering, and exploitation of mineral resources by both large-scale and small-scale miners and bush fires during the harmattan season (Kyere-Boateng and Mareck, 2021; Amponsa, 2002). There is little or no concerted effort to conserve these medicinal resources with the erroneous assumption that the plants will be available on a continuing basis (WHO, 1993). For most countries, especially developing countries that depend extensively on these plants, there is little or no attempt to catalogue the extent of medicinal plant loss and ascertain which medicinal plants are becoming rare or endangered. It is on record that about 15,000 species of medicinal plants are globally threatened largely due to commercial over-harvesting to meet the demand from drug manufacturers (Bentley, 2010). This has dire implications for the health and livelihood of many families. Therefore, there is the urgent need to bring important wild species of medicinal plants into cultivation as a biodiversity conservation strategy. It relieves harvesting pressure on rare and threatened species and assures the continuous supply of raw materials without endangering the survival of plant species (Chen et al., 2016). Despite the crusade for medicinal plant cultivation, certain constraints are evident. Medicinal plants are cultivated by smallholder farmers with lack of access to large swaths of land, quality propagation materials, non-existent market channels, and financial support. These must be addressed and the farmers provided incentives to be sensitive to biodiversity conservation through cultivation and sustainable harvesting methods (Nwafor et al., 2021).

For sustainable production of phytomedicines, it is also important to conduct value chain analysis of various medicinal plant species on a country-by-country basis. Previous analysis has shown that farmers who are the primary producers in the chain do not benefit as much as middlemen with better access to capital and information (Hishe et al., 2016). Majority of these producers are the rural poor who use the sale of medicinal plants as a source of income as well as means of diversifying household livelihoods. It has been suggested that the expansion of the fair trade scheme to include medicinal plants and not just agricultural food products will have a beneficial effect on both the incomes of farmers and the quality of materials produced. The fair trade movement is aimed at offering better deals for farmers by paying above the market rate for the commodity in question (Booker et al., 2012). In return, cultivators or farmers are expected to adhere to the fair trade policies on production and follow quality-driven requirements in key areas, particularly in the cultivation and collection stages. However, Farnsworth and Goodman (2008) noted that the scheme may be suitable in some countries than others. It is further argued that fair trade schemes tend to favour larger companies, and despite the presumed benefits, it has had little influence on the wages of the poorest workers or farmers (Booker et al., 2012).

Thus, it is evident that producers or cultivators of medicinal plants accrue limited direct benefits from participation in the value chain, and effort should be made to provide high incentives for cultivation as a biodiversity conservation tool and for improved livelihoods (Hishe et al., 2016). Efforts to improve the bargaining position of producers must address the multifaceted nature of the relationship between them and collectors or middlemen, wholesalers, and exporters. Integrating producers of medicinal plants into a broader, existing cooperative structure could improve their bargaining position and grant them access to small credits (Criss et al., 2006). The legalization of herbal medicine practice in many countries across the globe will further stimulate demand for high-quality medicinal plants and strengthen their value chain overall.

4.2 Cultivation of Medicinal Plants

Medicinal plants harvested from the wild are generally considered to be more efficacious than those that are cultivated. Wild-collected plants are an open resource without investment and could be free from pesticides (Chen et al., 2016). However, the practice risks extinction of ecotypes and species through overharvesting and recently climate change. Cultivation, on the other hand, relieves harvesting pressure on rare and threatened species. It is the only way to provide medicinal raw materials without further endangering the survival of rare, endangered, or over-exploited plant species (Raina et al., 2011; Hamilton, 2004). Moreover, it is an important means of biodiversity conservation. It is also the sustainable approach of ensuring the availability of a natural source of high-value industrial raw material for pharmaceutical, agrichemical, food, and cosmetic industries and opens new possibilities for higher level of gains for farmers with a significant scope for progress in rural economy (Chen et al., 2016). Cultivating medicinal plants does not deplete wild stocks, but rather, can replenish them and ensure sustainable supply. This is beacuse in many cases, the declining habitats of native plants can no longer supply the expanding market for medicinal plant products. Cultivation provides the opportunity to use new techniques to solve problems encountered in the production of medicinal plants, such as the presence of toxic components, pesticide contamination, low contents of active ingredients, and misidentification of plant species (Schipmann et al., 2005; Raina et al., 2011). Cultivation under controlled growth conditions can improve the yields of active compounds, ensures production stability, and prevents misidentification and adulteration (WHO, 2003).

Wild-collected plants normally vary in quality and composition because of environmental and genetic differences. This variation is much reduced by cultivation, and the production of raw materials of standard quality can be improved by the use of manure and fertilizers (Chen et al., 2016). Collection, drying, and storage are made easier, and drugs subject to control may be better regulated. The disadvantages include the substantial investment before and during production, infestation of plants under cultivation by pests and disease, and the difficulty in mimicking the natural environment for the cultivation of quality plant materials. Cultivation also narrows genetic diversity in the gene pool of wild populations and may have negative impact on ecosystems (Chang et al., 2011). The prevailing situation of medicinal plant cultivation differs from region to region, and therefore, an evaluation of current and future trends in production is warranted.

4.2.1 Current state of medicinal plant cultivation

Many countries have long traditions of cultivating medicinal plants. This is particularly true for countries in Asia where MPs play significant roles in ancient traditional medical systems such as Ayurveda of India, Unanic medicine (a Perso-Arabic traditional medicine as practiced in south and central Asia), and traditional Chinese medicine (Chapman and Chomchalow, 2003; WHO, 1993). They also generate income for the people of many Asian countries, who earn a living from selling collected materials from the forest or cultivate them on their lands. In most countries, only few farmers are presently engaged in the cultivation of medicinal plants.

Medicinal plant cultivation has grown in South-East Asia and China for the conservation of valuable species and generated income for the local people. However, only about 3.3% of most medicinal plants are cultivated, with the remaining proportion obtained from the wild. Presently, only few countries, such as China, India, Indonesia, Nepal, Thailand, and Vietnam, produce and commercialise medicinal and aromatic plants (Astutik et al., 2019). In these countries, cultivation is usually seen as an approach to improve local livelihoods. For example, in China, cultivated species contribute 10%–70% of the household income for the local people, whereas private farmlands in Nepal account for 20% of the household income (Phondani et al., 2018; Astutik et al., 2019). Medicinal plant production in some Asian and African countries is characterised by the following:

a. Subsistence cropping systems: Medicinal plants are grown by smallholder farmers in subsistence cropping systems. This includes the use of primitive cultivars grown in mixed cropping, with economic crops resulting in very low yield and quantities. In most cases, the farms are widely scattered, making collection or harvesting difficult for middlemen.
b. Poor-quality materials: This is mainly down to the use of low yielding varieties, poor cultural techniques and post-harvest handling.

c. Lack of integration: In many areas of the world where medicinal plants are grown as commercial crops, they are not integrated into farming systems as monoculture. They are normally cultivated among food crops that are considered primary crops as they bring income and basic foods to the farmers.

4.2.2 Constraints in production of medicinal plants

Despite the global outcry for the large-scale cultivation of medicinal plants to avert potential extinction through overharvesting, several challenges are still evident. Cultivation of MPs is affected by both biotic and abiotic factors (Chapman and Chomchalow, 2003).

4.2.2.1 Biotic

Most commercially cultivated medicinal plants are obtained from the wild, and these usually have low quality or yield. Genetic erosion and small germplasm collections in seed or gene banks or field gene banks means breeders have a difficult task. Several botanic gardens have been established for exhibition or tourist purposes without genetic diversity and not for the specific purpose of germplasm enhancement (Díez et al., 2018). Also, many of medicinal plants possess a long-life cycle, which often makes commercial cultivation costly and impractical. This is particularly true for plants whose useful parts are underground organs such as roots, tubers, and rhizomes such as Rauwolfia, Ginseng, Periwinkle, and Dioscorea species (Chapman and Chomchalow, 2003). Dioscorea, for example, is cultivated for its tubers rich in diosgenin, a steroid used as contraceptive, and will yield a maximum content of diosgenin only after five to six years of growth. Others have their active medicinal compounds in the stem bark which need to be stripped off, or in some cases, the entire plant cut down, resulting in plant loss or taking a long time for the next harvest. Additionally, medicinal plants cultivated as monocrops are susceptible to pest and diseases, and this affects the quality of the materials and cost of production (Chapman and Chomchalow, 2003; Gouvea et al., 2012).

4.2.2.2 Abiotic

Non-living factors affecting medicinal plant production include soil fertility, flood, drought, and climate change. MPs are cultivated as secondary crops likely to be grown in nutrient-deficient soils, resulting in poor yield and quality (Namdeo, 2018). Application of fertilizer may help improve the soil quality but has cost implications, which make it deterrent for poor rural farmers. Too much or too little water available for the growth of medicinal plants may result in stunted growth or even death. Improper light intensity and duration affects the production of MP. The growth of certain medicinal plants is severely affected when monocrop without the shade of big trees (WHO, 2003). However, for others such as the lemon grass (*Cymbopogon citratus*), they disappear if grown as an intercrop underneath big trees. Plant survival also depends on the optimum range of temperatures. Too low or too high temperatures are detrimental to the growth of medicinal plants. That is why plants introduced from the temperate zone often cannot do well in tropical Asia unless grown at high altitudes (Bita and Gerats, 2013; WHO, 2003).

4.2.2.3 Technological

Due to the lack of research and development, no technology for medicinal plant cultivation has been developed. Many farmers who grow medicinal plants lack the technical know-how on seed sowing, transplanting of seedlings, plant maintenance, harvesting, and collection practices. Poor harvesting methods result in deterioration in the quality of the materials. In most developing countries, facilities for research and development and services for medicinal plants are not available (Mohâ and Al-Uzaizi, 2016; Chapman and Chomchalow, 2003).

4.2.2.4 Socioeconomic

Most species of medicinal plants give low yields and their cultivation and harvesting are more labour-intensive. Many farmers would consider not growing MPs unless they compete with other economic crops in terms of income (Henle, 1993). Again, because the farmers do not have previous experience in growing medicinal plants, they prefer cultivating other traditional crops. In addition, there is often a lack of marketing channels for medicinal plants in most countries unlike economic crops. Also, there is no guarantee of a sustained market demand for MPs. Farmers cultivating MPs do not get any funding or incentive from the government or private sector (Chapman and Chomchalow, 2003). All these factors are important constraints that must be addressed to stimulate interest in cultivating medicinal plants.

4.3 Good agricultural practices for the cultivation of medicinal plants

Cultivation of medicinal plants requires careful planning and management. The conditions and duration of cultivation required vary depending on the quality of the medicinal plant materials required. If no scientific published or documented cultivation data are available, traditional methods of cultivation should be followed where feasible. Otherwise, a method should be developed through research. Good agricultural practices (GAP) in cultivation involving plant identification, propagation, maintenance, and harvesting should be followed (WHO, 2003)

4.3.1 Plant selection and authentication

Medicinal plant species specified in the national pharmacopoeia or authoritative document of the country could be considered for cultivation. Plant selection could also be due to economic value, demand, and whether the plant is endangered or overharvested. Others specified in the pharmacopoeia of other countries could be considered, and in the case of newly introduced medicinal plants, the species or botanical variety selected for cultivation should be identified and documented

According to the WHO guidelines on GAP, the botanical identity (genus, species, subspecies/variety, author, and family) of each medicinal plant under cultivation should be verified and recorded. If available, the local and English common names should also be recorded. Other relevant information, such as the cultivar name, ecotype, chemotype, and specific region or origin of material should be recorded as appropriate. Specimen should be submitted to a regional or national herbarium for identification. Where possible, a genetic pattern should be compared with that of an authentic specimen (WHO, 2003).

4.3.2 Propagation materials

It is important that the material for propagation (seeds, stems, tubers, rhizomes, etc.) are specified by the suppliers. The propagation parts of some important West African medicinal plants are listed (Table 1). All the necessary information relating to the identity, quality, and performance of the plant, including breeding history where necessary, should be provided. Propagation or planting materials should be of the appropriate quality and be as free as possible from contamination and diseases. Planting material should preferably be resistant or tolerant to biotic or abiotic factors.

Compared to other economic crops, medicinal plants have received much less attention in terms of genetic improvement. This is evident in the number of named cultivars used in commercial cultivation, which is surprisingly low. Lack of improved cultivars is due to the lack of germplasm conservation, facilities, and breeders and the lack of demand for large-scale cultivation of medicinal plants. To obtain excellent cultivars of medicinal plants that yield marketable standard products, some farmers obtain them from other countries as the easiest and less time-consuming approach. Secondly, the selection of desired genotypes can be made from existing variants. For example, in India, the selection of Rauvolfia serpentina RS-1 variety from a germplasm collection for cultivation afforded high yield of roots containing stable amount of reserpine, serpentine, and ajmalicine (Gupta, 1993). Equally important in the cultivation of medicinal plants is the site for propagation.

4.3.3 Geographical location or cultivation site

Medicinal plant materials derived from the same species can show significant differences in quality when cultivated at different sites, owing to the influence of soil, climate, and other factors. For example, the medicinal plant artichoke (*Cynara cardunculus var. scolymus*) cultivated in the mountainous Lào Cai province of Vietnam is found to contain high concentrations of the active ingredient cynarin than those grown in low-lying areas. Subsequently, the plant is cultivated in mountainous areas. The ecological and geographical variables affect the biosynthesis and hence the active constituents.

4.4 Climatic and ecological factors

The cultivation of medicinal plants may affect the ecological balance and genetic diversity of the flora and fauna in surrounding habitats. Growth of medicinal plants can also be affected by other plants, other living organisms, and human activities. All these factors should be considered when designing a medicinal plant cultivation scheme (WHO, 2003). Additionally, it is very important to assess the impact of cultivation on the biological and ecological balance of the region if a non-indigenous medicinal plant species is to be introduced, and this should be monitored over time where practical. Consideration should also be given to the climatic conditions, including the length of day, rainfall (water supply), and field temperature as these significantly influence the physical, chemical, and biological qualities of medicinal plants.

Table 1: West African medicinal plants and their mode of propagation (Amponsah, 2002)

Medicinal Plants	Medicinal Uses	Method of Propagation
Aframomum melegueta (Fruit)	Treatment of boils, rheumatism, and bone fractures	Rhizomes
Azadirachta indica (Leaves)	Treatment of ringworm, boils, fever, and hepatitis	Seeds
Balanites aegyptica (Bark)	Treatment of skin rashes and other ailments	Seeds
Bryophyllum calycinum (Leaves)	Treatment of whooping cough and whitlow	Stem
Cryptolepis sanguinolenta (Root)	Treatment of malaria and insomnia	Seeds
Clausena anisata (Leaves)	Treatment of arthritis	Seeds
Croton membranaceous (Bark)	Treating urinary retention and measles	Seeds
Diallium guinnensis (Root bark)	Treatment of cough	Seeds
Dioclea reflexa (Seed)	Treatment of asthma	Seeds
Hileria latifolia (Flowers)	Treatment of asthma	Stem
Khaya senegalensis (Bark)	Treating anaemia, arthritis, and malaria	Seeds
Ocimum gratissimum (Leaves)	Treatment of bacterial infections and diarrhoea	Seeds
Jatropha curcas (Leaves)	Treating wounds, convulsions, and fever	Seeds
Tetrapleura tetraptera (Bark)	Treating of gastric ulcer and dysentery	Seeds
Zingiber officinale (Rhizome)	Treatment of coughs, haemorrhoids, boils, and whooping cough	Rhizomes
Parquetina nigrescens (Leaves)	Treating gonorrhoea, jaundice, and rickets	Seeds
Starchytarpheta angustifolia (Leaves)	Treating asthma	Stem
Penianthus zenkeri (Roots)	Treating male sexual impotence, cough, and wounds	Seeds

Mormordica charantia (Leaves)	Treatment of hypertension, chicken pox, and cut wounds	Seeds
Monodora myristica (Seeds)	Treatment of anaemia, haemorrhoids, and sexual weakness	Seeds

4.4.1 Soil

Quality of soil is an important factor in medicinal plant cultivation. There is the need to evaluate the impact of past land uses, including the planting of previous crops and any applications of plant protection products. Risk of contamination because of pollution of the soil, air, or water by hazardous chemicals should be assessed. The soil for cultivation should contain appropriate amounts of nutrients, organic matter, and other elements to ensure optimal medicinal plant growth and quality. Optimal soil conditions, including soil type, drainage, moisture retention, fertility, and pH, will be dictated by the selected medicinal plant species and/or target medicinal plant part (WHO, 2003). The use of fertilizers may be necessary in some cases to obtain large yields of some medicinal plants. Fertilizer use in medicinal plant cultivation, if unregulated, often results in high biomass production but low concentration of active plant secondary metabolites. It is therefore necessary to ensure that correct types and quantities of fertilizers are used throughout. Consideration should be given to the application of organic fertilizers such as green manure and biofertilizers. Green manure includes animal droppings, composted plant, and animal remains or ploughing an herbaceous crop under and mixed with the soil while still green to enrich it (Namdeo, 2018). Green manure and cover crops prevent erosion and conserve soil nutrients. Animal manure should be thoroughly composted to meet safe sanitary standards of acceptable microbial limits.

Biologically active products or bacteria, algae, and fungi that are useful in bringing about soil nutrient enrichment (biofertilizers) may be considered. Human excreta must not be used as a fertilizer, owing to the potential presence of infectious microorganisms or parasites. All fertilizing agents should be applied sparingly and in accordance with the needs of the medicinal plant species and supporting capacity of the soil (Schippmann et al., 2006; Namdeo, 2018). Organic manure is benign to the environment and maintains biological processes of medicinal plants and ecological balance of habitats (Macilwain, 2004; Chen et al., 2016). Organic farming of medicinal plants affords materials with better quality and high productivity. As much as possible, agrochemicals should be excluded in organic farming.

4.4.2 Plant maintenance

Use of agrochemicals should be kept to a minimum and applied only when no alternative measures are available. Residues of agrochemicals contaminate medicinal plants and reduce their quality (Shaban et al., 2016; WHO, 2003). When necessary, only approved pesticides and herbicides should be applied at the minimum effective level. Only qualified staff using approved equipment should carry out pesticide and herbicide applications. All applications should be documented and international agreements on pesticide use and residues, such as the International Plant Protection Convention and Codex Alimentarius, adhered to (Namdeo, 2018). Integrated pest management should be followed where appropriate. Irrigation and drainage should be controlled and carried out in accordance with the needs of the individual medicinal plant species during its various stages of growth. Care should be exercised to ensure that the plants under cultivation are neither over- nor under-watered.

4.5 Social and economic impact of cultivating medicinal plants

The social impact of cultivation on local communities should be examined to prevent its negative effects on local livelihood. For example, it is not advisable for large swarths of land, hitherto used to cultivate food crops, to be used for medicinal plant cultivation on a large scale. In terms of local income-earning opportunities in rural communities, small-scale cultivation is often preferable to large-scale production. If large-scale medicinal plant cultivation is to be established, it is important for local communities to benefit directly from, for example, fair wages and equal employment opportunities (WHO, 2003).

The agreement between the Sapanapro Company and the Red Dao ethnic community of Sa Pa, Vietnam, highlights the socioeconomic impact of medicinal plant cultivation (Viet Lan, 2018). Sapanapro Company is a community enterprise set up to commercialise traditional bathing medicines of the Red Dao ethnic people. Their main products are bath medicines for women after pregnancy, which are based on the traditional knowledge of the Red Dao ethnic group. The company engages the Red Dao people themselves in the protection of their resources and traditional knowledge through sharing of accrued company benefits from commercialised products back to Red Dao communities. These benefits include monetary remunerations to the custodians of the traditional knowledge (bà mế in Vietnamese), and the company also pays the collectors of medicinal plants according to the value of the plant species collected. In addition, the company contributes a portion of the company profits to the communal development fund for the community's sociocultural activities. The model contributes both to the improvement of the local community's livelihood and to biodiversity conservation.

4.6 Medicinal plant value chains

According to Booker et al. (2012), a value chain (VC) describes the sequence of activities involved in the development of a product and the relationship between the various actors in the chain and on their implications for development. Value chain analysis has been used as a tool to understand the socioeconomic benefits, disadvantages, and risks for the various players along the chain. For example, in the VC analysis of non-timber forest products, by assessing the benefit to the primary producers, compared to the middlemen, wholesalers, and retailers, it was noted that farmers or gatherers only obtained a minimal share in the benefits of such products (Litvinoff and Madeley, 2007). This is also very common with economic crops.

The WHO estimates the global demand for medicinal plants at \$14 billion per annum (Booker et al., 2012). Such global value chain analysis has been criticised for its inability to explain consumption patterns and contribution to the understanding of general conditions of employment and production in poor countries, for example. Therefore, individual medicinal plants value chain analysis for specific regions give a better reflection of the impact on players in the chain. Although value chains have been widely used in various products, only a few studies have focused on medicinal plants and their products.

4.6.1 Actors of the medicinal plant value chains

The medicinal plant value chain is made up of the cultivators (producers), collectors, processors, wholesalers, exporters, and retailers (Chhabra, 2018). The upstream actors comprise producers and collectors who provide the basic raw materials and inputs for the other participants to function. The processors, wholesalers, retailers or traders, and exporters, who constitute the downstream actors, impart value to the products through processing and packaging, thereby enhancing their utilisation. The returns received by these actors constitute the total trade in medicinal plants (Booker et al., 2012).

4.6.1.1 Producers and collectors

This group consists of wildcrafters, cultivators, and plantation operators (Juliard et al., 2006). They are the first link and constitute the largest group in the chain, made up of farmers or members of the community where medicinal plants grow. Usually, the plants are collected from the wild or cultivation sites in rural communities. The collection and marketing of these medicinal plants from the wild is an important source of livelihood for many of the poor in developing countries (Chhabra, 2018; Booker et al., 2012). For example, it is on record that over three hundred thousand households are engaged in the collection of medicinal plants in Nepal, and these families are dependent on the income accrued from the sale of medicinal plants. The amounts paid for the plants are small but very significant for the survival of these families. In most cases, the collectors or producers are exploited by the processors, wholesalers, exporters, and retailers (Morser, 2010). The bargaining power of the producers appears to be weak compared to the other actors who can exert control on the price paid for the medicinal plants (Chabbra, 2018). The main constraint of this group is its marginalization and general lack of information. There is therefore the need to explore opportunities for improving the livelihood of collectors by establishing communication links, improving their power to negotiate through producer associations, providing stable markets, or creating opportunities for local-level value-added processing (Hishe et al., 2016). Producers are aware of biodiversity issues but have limited knowledge and incentives to apply conservation practices at the ground level (Juliard, 2006).

4.6.1.2 Collectors and processors

Middlemen or collectors harvest plant materials from wildcrafters and cultivators and sell to processors. Processors convert vegetative material into bulk or consumer-grade products. They rid the plant materials off organic and other foreign matter. Sometimes they dry, mill, extract or prepare concentrates, package, and store them in readiness for the market (Hishe et al., 2016). Processing of the medicinal plant products adds value to them, thereby allowing them to charge higher prices for the same. Sometimes middlemen intervene as facilitators due to lack of efficient linkages between the cultivators and the wholesalers and the retailers and the pharmaceutical or herbal industries (Chhabra, 2018). This group usually does not care about biodiversity conservation issues.

4.6.1.3 Wholesalers and exporters

These well-organised actors provide an effective link between cultivators, processors, and middlemen in the medicinal plant industry. The wholesalers are the distributors of the medicinal plant products to the ultimate markets, and they carry out the work through a complex network of agents and retailers. They are usually large firms with extensive financial capacities (Taghouti et al., 2022). They provide the producers and collectors with valuable information regarding consumer demand for the medicinal plant products in the domestic as well as in the international markets. The wholesalers prepare the plant materials (raw or processed) to be given to the consumers, the final recipient of the product. Sometimes, collectors produce commercial end products that go directly to the final consumer (Chhabra, 2018; Taghouti et al., 2022).

4.6.1.4 Pharmaceutical and cosmetic industries

The pharmaceutical and herbal industries are important secondary actors in the value chain of medicinal plants. They use the medicinal plants or their extracts as input materials and then add value to them through processing to afford the final product, which is consumer friendly. These industries are the prime example of the value addition made to medicinal plants.

4.7 A snapshot of few studies in medicinal plant value chains

While there is a relative abundance of reports of the value chain of a range of food products (Menon, 2008), only a limited number of studies on herbal medicines exist. Most of these reports are dominated by indigenous medicinal plants of Asia with almost no reports from West Africa (Hishe et al., 2016).

4.7.1 Lessons from the cultivation of kutki (Picrorhiza kurrooa) in Northern India

Alam and Belt (2009) reported on a project in Uttarakhand, Northern India, aimed at the cultivation of a medicinal plant, Kutki (*Picrorhiza kurroa Royle ex Benth*), with the objective of providing financial and social benefits to the farmers and helping preserve wild species. International partners from Europe were expected to benefit from having a secure supply of the plant from a fully traceable source. The Uttarakhand project reportedly produced disappointing results because planting materials were of poor quality. Planting was done on small poorly irrigated lands, and the emergence of apples as a profitable cash crop eventually led to farmers switching from kutki to apples.

4.7.2 Medicinal plant production and value chain in Bangladesh

In Bangladesh, where primary and wholesale secondary markets for medicinal plants are dominated by middlemen, Shahidulla and Haque (2010) reported that a vertically integrated value chain benefited producers and processors at the beginning of the value chain. This approach enabled primary producers to become active participants in the process. It removed market access barriers, resulting in better commercialisation of products, and was attractive to companies as they could have greater control over quality and supply. It was suggested that to sustain growth in medicinal plant production, a fair distribution of the gross margin to the primary producers is necessary. In the value chain system examined, it was found that downstream buyers, especially manufacturers and consumers, pay most of their money for middlemen's value addition opportunistic pricing due to inherent weaknesses in the chain. A vertically integrated chain, with only producers and processors as commercial actors and nongovernmental organisations as promoters, creates an economically robust system which benefits the many rather than the few (Shahidulla and Haque, 2010; Volenzo and Odiyo, 2020).

4.7.3 Medicinal plants for biodiversity conservation

Several conservation approaches have been proposed, providing both in situ and ex situ conservation. Natural reserves and wild nurseries are typical examples to retain the medicinal efficacy of plants in their natural habitats, while botanic gardens and seed banks are important paradigms for ex situ conservation and future replanting. The geographic distribution and biological characteristics of medicinal plants must be known to guide whether species conservation should take place in nature or in a nursery (Chen et al., 2016). There are two methods for the conservation of plant genetic resources, namely, *in-situ* and *ex-situ* conservation.

4.7.3.1 In situ

In situ conservation means the conservation of ecosystems and natural habitats and the maintenance and recovery of viable populations of species in their natural surroundings. This includes conservation of forest through protected areas like national parks and wildlife sanctuaries (Bentley, 2010). This type of conservation is achieved both by setting up areas as nature reserves and wild nurseries and by ensuring that as many wild species as possible can still survive in managed habitats such as farms and plantation forests (Hamilton, 2004).

4.7.3.2 Wild nurseries

It is impossible to designate every natural wild plant habitat as a protected area, owing to cost considerations and competing land uses. A wild nursery is established for species-oriented cultivating and domesticating of endangered medicinal plants in a protected area, natural habitat, or a place that is only a short distance from where the plants naturally grow. Although the populations of many wild species are under heavy pressure because of overexploitation, habitat degradation and invasive species, wild nurseries can provide an effective approach for *in situ* conservation of medicinal plants that are endemic, endangered, and in-demand (Li and Chen, 2007; (Soule et al., 2005).

4.7.3.3 Nature reserves

Nature reserves are protected areas of important wild resources created to preserve and restore biodiversity (Rodriguez et al., 2007). According to Liu et al. (2001), more than 12,700 protected areas have been established, accounting for 13.2 million km^2 or 8.81% of the earth's land surface. Conserving medicinal plants by protecting key natural habitats requires assessing the contributions and ecosystem functions of individual habitats.

4.7.3.4 Ex situ

This is the conservation of components of biological diversity outside their natural habitats. In ex situ conservation, endangered or rare species are transferred from their natural habitats to protected areas equipped for their protection and preservation. Priority for ex situ conservation should be given to species whose habitats may have been destroyed or cannot be safeguarded. Ex situ conservation techniques include botanic gardens and seed banks (Chen et al., 2016)

4.7.3.5 Botanic gardens

Botanic gardens play an important role in ex situ conservation, and they can maintain the ecosystems to enhance the survival of rare and endangered plant species. They involve a wide variety of plant species grown together under common conditions and often contain taxonomically and ecologically diverse flora (Primack and Miller-Rushing, 2009). They can play a further role in medicinal plant conservation through the development of propagation and cultivation protocols, as well as undertaking programmes of domestication and variety breeding (Chen et al., 2016).

4.7.3.6 Seed banks

Seed banks offer a better way of storing the genetic diversity of many medicinal plants ex situ than through botanic gardens, and are recommended to help preserve the biological and genetic diversity of wild plant species. The most noteworthy seed bank is the Millennium Seed Bank Project at the Royal Botanic Gardens in Britain (Schoen and Brown, 2001). Seed banks allow relatively rapid access to plant samples for the evaluation of their properties, providing helpful information for conserving the remaining natural populations. They should also be used to bulk up populations of depleted or even locally extinct plants for restocking in nature (Li and Pritchard, 2009).

4.8 Conclusion

A deliberate, environmentally conservative medicinal plant cultivation is needed at a time when there is an accelerated depletion of medicinal plant resources about 1000 times higher than their expected natural extinction rate due to high global demand, increase in human population, and unsustainable harvesting practices. Cultivation of medicinal plants is a biodiversity conservation strategy that relieves harvesting pressure on the threatened species, assures the continuous supply of raw materials, and serves as a source of livelihood for many rural poor in developing countries at the base of the value chain. High-quality medicinal plants can be produced by following the guidelines on good agriculture and collection practices recommended by the World Health Organization. Key points in the cultivation include carefully selecting the plant species and obtaining high-quality propagation materials. Selection of the cultivation site should consider previous land use, soil type, and environmental and climatic conditions.

Smallholder farmers who dominate the cultivation of medicinal plants are constrained by poor-quality propagation materials characterised by low yields and long life cycles. Lack of information on new technologies in agronomy, unavailability of market channels, and low earnings from the sale of medicinal plants also deter farmers from engaging in their cultivation. In the value chain analysis of medicinal plants, the cultivators are also exploited by middlemen, wholesalers, and exporters, thereby necessitating the need to improve their power to negotiate through producer associations, availability of small credits, and provision of stable markets or by creating opportunities for local level value-added processing and establishing communication links.

Value chain analysis of medicinal plants, unlike food crops, is not widespread. There is a need to conduct research into the value chain of African medicinal plants and their products to ascertain their socioeconomic benefits, among other things. It is also recommended that both in situ and ex situ conservation methods, such as wild nurseries, nature reserves, botanic gardens, and seed banks, be explored to preserve endangered species and ensure a continuous supply of propagation materials. These conservation methods are environmentally friendly and easily adaptable by smallholder farmers in developing countries to safeguard the earth's rich medicinal plant flora.

References

1. Alam G, Belt J (2009). Developing a medicinal plant value chain: Lessons from an initiative to cultivate Kutki (Picrorhiza kurrooa) in Northern India. KIT Working Papers Series, (WPS. C5). https://www.kit.nl/wp-content/uploads/2018/08/1632 Developing-a-medicinal-plant-value-chain WPS-C5.pdf

2. Amponsah K, Crensil O, Odamtten GT, Ofusohene-Djan W (2002). Manual for the propagation and cultivation of medicinal plants of Ghana. https://www.bgci.org/files/Africa/pdfs/manual for the propagation and cultivation of medicinal plants of ghana.pdf

3. Astutik S, Pretzsch J, Ndzifon Kimengsi J (2019). Asian medicinal plants' production and utilization potentials: A review. Sustainability, 11(19):5483.

4. Bentley R (2010). Medicinal plants. London: Domville-Fife Press, 23–46.

5. Bita CE, Gerats T (2013). Plant tolerance to high temperature in a changing environment: scientific fundamentals and production of heat stress-tolerant crops. Frontiers in Plant Science, 4:273.

6. Booker A, Johnston D, Heinrich M (2012). Value chains of herbal medicines—Research needs and key challenges in the context of ethnopharmacology. Journal of Ethnopharmacology, 140(3):624–33.

7. Brijesh KS (2011). Rauwolfia: cultivation and collection. *Biotech Articles Web site.* http://www.biotecharticles.com/Agriculture-Article/Rauwolfia-Cultivationand-Collection-892.html.

8. Chang LI, Hua YU, Shi-Lin CH (2011). Framework for sustainable use of medicinal plants in China. Plant Diversity, 33(01): 65.

9. Chapman K, Chomchalow N (2003). Production of medicinal plants in Asia. InIII WOCMAP Congress on Medicinal and Aromatic Plants-Volume 5: Quality, Efficacy, Safety, Processing and Trade in Medicinal, 679: 45–59.

10. Chen SL, Yu H, Luo HM, Wu Q, Li CF, Steinmetz A (2016). Conservation and sustainable use of medicinal plants: problems, progress, and prospects. Chinese medicine, 11(1):1–0.

11. Chhabra T (2018). Value Chain analysis for medicinal plant-based products in India: Case study of Uttarakhand. Archives of Organic and Inorganic Chemical Sciences, 4:449–57.

12. Dias DA, Urban S, Roessner U (2012). A historical overview of natural products in drug discovery. Metabolites, 2(2):303–36.

13. Díez MJ, De la Rosa L, Martín I, Guasch L, Cartea ME, Mallor C, Casals J, Simó J, Rivera A, Anastasio G, Prohens J (2018). Plant genebanks: Present situation and proposals for their improvement. the case of the Spanish network. Frontiers in Plant Science, 9:1794.

14. Dorm-Adzobu C (1982). Impact of utilization of natural resources on forest and wooded savanna ecosystems in rural Ghana. Environmental Conservation, 9(2):157–62.

15. Farnworth C, Goodman M (2008). Growing ethical networks: The Fair-Trade market for raw and processed agricultural products. Washington, DC: World Bank. https://documents1.worldbank.org/curated/en/893431468045069908/pdf/413600Growing0Ethical0Networks01PUBLIC1.pdf

16. Gouvea DR, Gobbo-Neto L, Lopes NP (2012). The influence of biotic and abiotic factors on the production of secondary metabolites in medicinal plants. Plant Bioactives and Drug Discovery: principles, practice, and perspectives, 17:419.

17. Hamilton AC (2004). Medicinal plants, conservation and livelihoods. Biodiversity and Conservation, 13(8):1477–517.

18. Havens K, Vitt P, Maunder M, Guerrant EO, Dixon K (2006). Ex situ plant conservation and beyond. BioScience, 56(6):525–31.

19. Henle HV (1993). Socio-economic aspects of medicinal and aromatic plant production in Asia. RAPA Publication (FAO), 1993/19.

20. Hishe M, Asfaw Z, Giday M (2016). Review on value chain analysis of medicinal plants and the associated challenges. Journal of Medicinal Plants Studies, 4(3):45–55.

21. Jamshidi-Kia F, Lorigooini Z, Amini-Khoei H (2018). Medicinal plants: Past history and future perspective. Journal of Herbmed Pharmacology, 7(1).

22. Juliard C, Benjamin C, Sassanpour M, Ratovonomenjanahry A, Ravohitrarivo P (2006). Madagascar Aromatic and Medicinal Plant Value Chain Analysis. Combining the value chain approach and nature, health, wealth and power frameworks. microREPORT 70. https://pdf.usaid.gov/pdf_docs/Pnadh969.pdf

23. Kyere-Boateng R, Marek MV (2021). Analysis of the social-ecological causes of deforestation and forest degradation in Ghana: Application of the DPSIR framework. Forests, 12(4):409.

24. Laith MR, Sammar FA (2016). Farmer's knowledge level and training needs toward the production and conservation of medicinal herbal plants in Jordan. Journal of Medicinal Plants Research, 10(24):351–9.

25. Li DZ, Pritchard HW (2009). The science and economics of ex situ plant conservation. Trends in Plant Science, 14(11):614–21.

26. Li XW, Chen SL (2007). Conspectus of ecophysiological study on medicinal plant in wild nursery. Zhongguo Zhong yao za zhi= Zhongguo zhongyao zazhi– China Journal of Chinese Materia Medica, 32(14):1388–92.

27. Litvinoff M and Madeley J (2007). 10 reasons to buy fairtrade. Available online at http://www.globaldimension.org.uk/docs/10 Reasons to Buy Fair Trade.pdf.

28. Liu J, Linderman M, Ouyang Z, An L, Yang J, Zhang H (2001). Ecological degradation in protected areas: the case of Wolong Nature Reserve for giant pandas. Science, 292(5514):98–101.

29. Macilwain C (2004). Organic: is it the future of farming? Nature, 428(6985):792–4.

30. Morser A (2010). A Bitter Cup: Exploitation of tea pickers in India and Kenya by British Sumermarkets. War on want report. Available online at http://www.waronwant.org/downloads/7128.1 WOW Tea Report prf9.pdf.

31. Namdeo AG (2018). Cultivation of medicinal and aromatic plants. In Natural Products and Drug Discovery, Pages 525–553, Elsevier.
32. Nwafor I, Nwafor C, Manduna I (2021). Constraints to cultivation of medicinal plants by smallholder farmers in South Africa. Horticulturae, 7(12):531.
33. Phondani PC, Bhatt ID, Negi VS, Kothyari BP, Bhatt A, Maikhuri RK (2016). Promoting medicinal plants cultivation as a tool for biodiversity conservation and livelihood enhancement in Indian Himalaya. Journal of Asia-Pacific Biodiversity, 9(1):39–46.
34. Primack RB, Miller-Rushing AJ (2009). The role of botanical gardens in climate change research. New Phytologist, 182(2):303–13.
35. Raina R, Chand R, Sharma YP (2011). Conservation strategies of some important medicinal plants. International Journal of Medicinal and Aromatic Plants, 1:342–7.
36. Rigby D, Cáceres D (2001). Organic farming and the sustainability of agricultural systems. Agricultural Systems, 68(1):21–40.
37. Rodríguez JP, Brotons L, Bustamante J, Seoane J (2007). The application of predictive modelling of species distribution to biodiversity conservation. Diversity and Distributions, 1:243–51.
38. Rousan LM and Al-Uzaizi SF (2016). Farmer's knowledge level and training needs toward the production and conservation of medicinal herbal plants in Jordan. Journal of Medicinal Plants Research, 10(24): 351–359.
39. Schipmann U, Leaman DJ, Cunningham AB, Walter S (2003). Impact of cultivation and collection on the conservation of medicinal plants: global trends and issues. InIII WOCMAP Congress on Medicinal and Aromatic Plants-Volume 2: Conservation, Cultivation and Sustainable Use of Medicinal and *676:* 31–44.
40. Schippmann U, Leaman DJ, Cunningham AB (2002). Impact of cultivation and gathering of medicinal plants on biodiversity: global trends and issues. Biodiversity and the ecosystem approach in agriculture, forestry and fisheries. https://www.researchgate.net/profile/Uwe-Schippmann/publication/265157471_Impact_of_Cultivation_and_Gathering_of_Medicinal_Plants

on Biodiversity Global Trends and Issues/links/553f24b60 cf294deef7193d9/Impact-of-Cultivation-and-Gathering-of-Medicinal-Plants-on-Biodiversity-Global-Trends-and-Issues.pdf

41. Schippmann UW, Leaman D, Cunningham AB (2006). A comparison of cultivation and wild collection of medicinal and aromatic plants under sustainability aspects. Frontis. 1:75–95.

42. Schoen DJ, Brown AH (2001). The conservation of wild plant species in seed banks: attention to both taxonomic coverage and population biology will improve the role of seed banks as conservation tools. BioScience, 51(11):960–6.

43. Shaban NS, Abdou KA, Hassan NE (2016). Impact of toxic heavy metals and pesticide residues in herbal products. Beni-suef University Journal of Basic and Applied Sciences, 5(1):102–6.

44. Shahidullah AK, Haque CE (2010). Linking medicinal plant production with livelihood enhancement in Bangladesh: Implications of a vertically integrated value chain. Journal of Transdisciplinary Environmental Studies 9(2):1.

45. Soulé ME, Estes JA, Miller B, Honnold DL (2005). Strongly interacting species: conservation policy, management, and ethics. BioScience, 55(2):168–76.

46. Taghouti I, Cristobal R, Brenko A, Stara K, Markos N, Chapelet B, Hamrouni L, Buršić D, Bonet JA (2022). The Market Evolution of Medicinal and Aromatic Plants: A Global Supply Chain Analysis and an Application of the Delphi Method in the Mediterranean Area. Forests, 13(5):808.

47. UN comtrade (2018). "Export-Import Value of Products Reported in Code HS1211," Trade Statistics Database Reported in Code H1211. Department of Economics and Social Affairs, Statistics Division, http://comtrade.un.org/data/

48. Van Andel T, Myren B, Van Onselen S (2012). Ghana's herbal market. Journal of Ethnopharmacology, 140(2):368–78.

49. Veeresham C (2012). Natural products derived from plants as a source of drugs. Journal of Advanced Pharmaceutical Technology & Research, 3(4):200.

50. Viet Lan N (2018). Preserving tradition for sustainable development. UNDP report. https://vietnam.un.org/en/21374-preserving-tradition-sustainable-development

51. Volenzo T, Odiyo J (2020). Integrating endemic medicinal plants into the global value chains: the ecological degradation challenges and opportunities. Heliyon 1;6(9):e04970. doi: 10.1016/j. heliyon.2020.e04970

52. Wachtel-Galor S and Benzie, IF (2011). Herbal Medicine: An Introduction to Its History, Usage, Regulation, Current Trends, and Research Needs. In: Benzie IFF, Wachtel-Galor S, editors. Herbal Medicine: Biomolecular and Clinical Aspects. 2nd edition. Boca Raton (FL): CRC Press/Taylor & Francis.

53. World Health Organization (2003). WHO guidelines on good agricultural and collection practices (GACP) for medicinal plants. World Health Organization. https://apps.who.int/iris/handle/10665/42783

54. World Health Organization, (1993). Guidelines on the conservation of medicinal plants. Gland: International Union for Conservation of Nature and Natural Resources. https://apps.who.int/iris/handle/10665/41651

5

Use of Good Manufacturing Practice in Small-Scale Phytomedicines Production: Principles and Practice

Doris Kumadoh[1,3], Mavis Boakye-Yiadom[2], Mary-Ann Archer[4], Genevieve Yeboah[1], Michael Odoi Kyene[1], Emmanuel Adase[3]

[1]*Department of Pharmaceutics and Quality Control, Centre for Plant Medicine Research, Mampong-Akuapem, Ghana*
[2]*Department of Clinical Research, Centre for Plant Medicine Research, Mampong-Akuapem, Ghana*
[3]*Production Department, Centre for Plant Medicine Research, Mampong-Akuapem, Ghana*
[4]*Department of Pharmaceutics, University of Cape Coast, Cape Coast, Ghana*

5.1 Introduction

Small and medium-scale enterprises (SMEs) have been recognised as vehicles for the achievement of growth objectives in Africa. It is estimated that SMEs provide approximately 80% of job opportunities across the continent, with about 44 million micro, small, and medium enterprises in sub-Saharan Africa alone (Runde et al., 2021). Small-scale phytomedicine production is increasing rapidly in Africa, and there is a need for these setups to follow good manufacturing practices (GMP) to ensure that products are regularly produced and guided to the quality standards applicable to their proposed use and as required by regulatory authorities for marketing approval. In ancient times, herbalists were known for manufacturing products for immediate use. These preparations were normally done at the premise of the traditional medicine practitioner/herbalist and guidance given on their usage for a period depending on the disease condition(s) being treated (Abubakar and Haque, 2020).

Currently, the need for phytomedicines on a relatively larger scale has led to the establishment of units meant to facilitate production on a reasonably larger scale suitable for long-term use. Products manufactured include liquid dosage forms such as decoctions, infusions, syrups, aromatic waters, tinctures, liquid inhalations, and medicated oils; semisolid dosage forms such as ointments, creams, and salves; and solid dosage forms such as teas, pills, granules, dry plant and extract powders, capsules, tablets, dry powder inhalations, lozenges, plasters, and patches (Kumadoh and Ofori-Kwakye, 2017; World Health Organization, 2018; Ozioma et al., 2019; Asare et al., 2021).

Some herbalists have ventured into small-scale manufacturing in an attempt to serve a relatively larger number of clients for a much longer period. A trend of an increase in the production of herbal preparations consisting mostly of complex mixtures of several medicinal plants and sold in numerous retail outlets, including supermarkets and pharmacies, has been observed in current African traditional medicine practice. This trend may be due to increasingly urbanised societies that rely largely on herbal products, but do not have the time or resources to produce them (Ndhlala et al., 2011).

The World Health Organization (WHO) defines phytomedicines (herbal medicines) as preparations, including herbs, herbal materials, herbal preparations, and finished herbal products whose active ingredients include parts of plants, other plant materials, or their combinations (World Health Organization, 2019). In some countries, herbal medicines may contain, by tradition, natural organic or inorganic active ingredients that are not of plant origin, such as animal and mineral materials, fungi, algae, and lichens, among others (World Health Organization, 2019). In the context of this chapter, however, this latter definition does not apply as reference is specifically made to phytomedicines which are solely plant based.

The application of good manufacturing practices (GMPs) has become a necessary tool for ensuring that the quality and safety of herbal medicines produced by these small-scale phytomedicine manufacturers are not compromised to cause harm to consumers. GMP is a quality assurance system that aims at ensuring that medicinal products are always manufactured and controlled to the required quality and marketing authorisation standards. GMP mainly aims at decreasing risks of contamination and cross-contamination (mix-ups or confusion) that cannot be controlled by testing the product (World Health Organization, 2007; World Health Organization, 2019; European Commission, 2017; South Africa Health Products Regulatory Authority, 2019).

This chapter aims at exploring the good manufacturing principles and practices in small-scale phytomedicine production in Africa, to guide small-scale phytomedicine manufacturers to produce products that are of good quality and safe for consumers without posing health risks. This work may also aid regulatory authorities in various African countries to sequence, review, and/or add to existing regulations for small-scale phytomedicine producers.

5.2 Small-scale phytomedicine production in Africa

Small-scale phytomedicine production refers to the manufacturing of phytomedicines in industries with the help of relatively smaller machines and a few employees. The industries ensure easy management of the sector, and direct relationships between the managers, employees, and customers are strengthened to enhance the flow of production and marketing. The aim is to produce phytomedicines that meet the required standards, with the aid of relatively smaller machines and a few workers and employees (normally between ten to hundred) (Natako, 2006). In Africa, the production of phytomedicines on a small scale has been done by traditional medicine practitioners and traditional healers at their homes (Natako, 2006). The use of clay pots, cooking pots, vessels, and buckets, among others, as equipment by some traditional healers for the production of phytomedicine is common. However, some small-scale phytomedicine industries currently use modern equipment for manufacturing.

5.3 Good manufacturing practices for phytomedicines

The composition of phytomedicines, consisting of phytochemical constituents and bioactive compounds, makes good manufacturing practices an important tool to ensure their quality and safety. When small-scale phytomedicine manufacturers implement GMPs, the end product is expected to be of good quality, efficacacious, and safe (US Food and Drug Administration, 2020; Kwesiga et al., 2021).

In the application of GMPs, standards are clearly defined for manufacturing activity and quality control processes. General considerations that ensure that defined processes lead to expected quality outcomes include:

- a clear definition of all processes,
- validation of the processes,
- review and documentation of all processes,
- engagement of qualified and trained personnel,
- use of appropriate premises and equipment,
- use of suitable materials (raw, packaging, etc.),

- application of required guidelines and legal regimes for contract production and testing where applicable, and
- responding promptly to product complaints and defects (https://www.who.int/teams/health-product-and-policy-standards/standards-and-specifications/gmp).

It is necessary to define the first critical step where GMPs are to be applied to give an appropriate flow of processes, which will subsequently lead to desired quality outcomes (Food and Drugs Authority, 2018; WHO, 2018). The element of GMP emphasises that manufacturing procedures or processes are fully clear before implementation. In practice, these include using good raw plant materials, proper premises, trained personnel, appropriate equipment, proper storage facilities, approved procedures, documentation or record keeping, and transport services (World Health Organization, 2018). In Ghana, the Food and Drugs Authority (FDA), which is responsible for the certification of phytomedicines sold on the Ghanaian market has adopted practically all the elements of GMP as outlined by the WHO for the manufacturing of phytomedicines. These include application of quality control systems, raw materials handling and usage, manufacturing operations, specifications for products, standard operating procedures, sanitation and hygiene, equipment handling, trained personnel, documentation, production area, medical screening of personnel, product recalls, retained samples, specification, complaints, labels, and packaging (Food and Drugs Authority, 2018). Most African countries also rely on the same WHO guidelines for GMP for the manufacturing of herbal medicines (Gouveia et al., 2015; Food and Drugs Authority, 2018).

5.4 Definitions

The following definitions may apply in this write-up:

5.4.1 Medicinal plants

Medicinal plants consist of wild or cultivated plants used for medicinal purposes (World Health Organization, 2019).

5.4.2 Herbs

Herbs include crude plant materials such as flowers, leaves, seeds, fruits, roots, stem bark, rhizomes, or other plant parts, which may be whole, powdered, or fragmented (World Health Organization, 2019).

5.4.3 Herbal materials

In addition to herbs, herbal materials include gums, fresh juices, essential oils, fixed oils, resins, and dry powders of herbs. In some countries, these materials may be processed by various local procedures, such as roasting, steaming, or stir-baking with alcoholic beverages, honey, or other plant materials (Gouveia et al., 2015; World Health Organization, 2019).

5.4.4 Herbal preparations

Herbal preparations are the basis for finished herbal products and may consist of powdered or comminuted herbal materials, tinctures or extracts, and fatty oils. Herbal preparations are normally produced by extraction, concentration, fractionation, purification, or other biological or physical processes such as heating or steeping in honey and/or alcoholic beverages or other materials (Gouveia et al., 2015; World Health Organization, 2007).

5.4.5 Finished herbal products

Finished herbal products consist of one or more herbal preparations made from one or more herbs. Products consisting of different plant materials are called mixture herbal products. In addition to the active ingredients, finished herbal products and mixture herbal products may contain excipients. However, finished products or mixture herbal products to which chemically defined active substances have been added, including synthetic compounds and/or isolated constituents from herbal materials, are not considered 'herbal' (Gouveia et al., 2015; World Health Organization, 2007).

5.4.6 Excipients

Excipients are inert substances that are included in pharmaceutical dosage forms (liquid dosage form, solid dosage form, and semisolid dosage forms) not for their therapeutic action, but to help the manufacturing process to significantly enhance stability and protect aesthetics, performance, patient acceptability, or patient compliance. They also aid in product identification and enhance the overall safety or function of the product during storage or use. The excipient must be validated and standardised. It can be categorised in different ways, such as the route of administration (oral, topical, parental, and other excipients), origin (organic chemicals and inorganic chemicals), and functionality (Alison and Beverley, 2011; Martindale, 2011).

5.4.7 Standard operating procedures (SOPs)

Standard operating procedures may be defined as written procedures prescribed for repetitive use as a practice, in accordance with the agreed-on specifications aimed at obtaining a desired outcome. SOPs help in the implementation of GMP when they are carefully followed, monitored, and evaluated with time (Al et al., 2012; Shukla, 2017; Centre for Plant Medicine Research, 2022; World Health Organization, 2018). An example of an SOP is presented in Annex 1: Sample SOP: SOP for personal hygiene and discipline.

5.5 Control of starting materials

Controlling starting materials is a vital step in phytomedicine manufacturing. If starting materials are of poor quality, the quality of the finished product will be adversely affected (Djordjevic, 2017; World Health Organization, 2018).

5.5.1 Control of herbal (plant) materials

Starting materials include raw plant materials, which serve as the active ingredient in the preparation. These plant materials may be obtained from the wild or cultivated (Guo et al., 2022). The WHO recommends that for each plant material used, the family and botanical name according to the binomial system (genus, species, variety, and authority) must be known and used. Vernacular names and therapeutic uses in the country of origin may be added and documented appropriately (World Health Organization, 2019). However, in most African countries, the herbalist may know the local names of the plants used, but not the botanical or scientific names. Mandated plant medicine institutions and regulatory agencies must therefore make a conscious effort to help phytomedicine producers name their plant materials used scientifically. This will go a long way towards helping producers in accurately identifying the starting herbal material. In Ghana, some effort has gone into linking local names in the various languages of plants to their botanical names (Boadu and Asase, 2017).

Another important aspect of the plant materials used is the source (i.e., country, region, town, city, or village of origin), and whether it was collected from the wild or cultivated, the procedures used in collection, and the part of the plant used. The producer must also find out if there is a likelihood of contamination with pesticides, chemicals, and heavy metals from the collection area (Shippmann et al., 2006; World Health Organization, 2019).

During collection and/or harvesting of plant materials, it should be ensured that

- medicinal plants are collected when matured and deemed to contain a maximum of active constituents; and
- plants are collected and/or harvested at the right time. The time of collection depends on the plant part or nature of the plant.

Some recommended periods for collection are as follows:

- Leaves and flowering tops may be collected at the time of flowering and before fruits and seeds mature.
- Flowers may be collected just before they are fully expanded.

- Leaves and flowers may be collected in dry weather.
- Fruits may be collected before or after the ripening period.
- Seeds are collected when fully mature, when most of them have ripened.
- Barks are normally collected at the end of the rainy season after a period of damp weather.
- Roots, rhizomes, and other underground organs are normally collected at the end of vegetative processes when organs contain a maximum of food reserve. They should be freed from sand and other earthy or vegetable extraneous materials (Nishteswar and Tavhare, 2014; Porwal et al., 2020; World Health Organization, 2019).

Phytomedicine producers should be encouraged to take photographs of plant materials used, in addition to keeping reference samples of materials (most practically in dried form to help in identification when necessary). This will help in visual identification, which is most applicable to small-scale phytomedicine producers in most African countries.

During harvesting, care must be taken to ensure plant materials are not contaminated with foreign matter, such as other parts of the same plant, parts of other plants, or non-plant materials. Plant materials must be carefully handled with the removal of all foreign matter before further processing, dried where applicable according to defined criteria, and stored appropriately in well-controlled and designated areas with appropriate conditions (Pandey, 2017; Porwal et al., 2020; World Health Organization, 2019).

Quality specifications for each plant material used should be clearly documented and defined. When plant materials do not comply with quality specifications, rules for their rejection, storage, and disposal should also be defined (European Commission, 2017; World Health Organization, 2019).

5.5.2 Control of non-plant-based starting materials

Control of non-plant-based starting materials such as excipients and primary or printed packaging materials, should be done using specifications from pharmacopeial monographs where required. These materials should be kept at places where they will not be contaminated. Primary packaging materials such as bottles and containers, where applicable, must be cleaned using established protocols before use as they come into direct contact with the product, hence their likelihood of introducing contamination into products (Gouveia et al., 2015).

5.6 Sanitation and hygiene

5.6.1 Sanitation

Adherence to simple sanitation rules may help to prevent contamination arising from poor sanitation. There should be regular (at least once daily) disposal of accumulated waste in waste bins. Waste bins must be available at needed points on the production premises, kept clean, and clearly marked to prevent their use in undesignated areas. Waste derived from the manufacturing facility should also be disposed of regularly to decrease the risk of contamination during manufacturing (Gouveia et al., 2015; European Commission, 2017; Food and Drugs Authority, 2018).

5.6.2 Personal hygiene

Personal hygiene is a key aspect of GMP. Producers must ensure that personnel maintain a high standard of personal hygiene (European Commission, 2017; Food and Drugs Authority, 2018). Simple hygiene steps, such as keeping short and clean nails, hair, and beard; regular washing of hands; and keeping the whole body clean in general, go a long way towards significantly reducing the risk of contamination of herbal products. Personnel with infectious diseases should not be allowed to work with herbal materials until they have been treated.

Written procedures listing the basic hygiene requirements should be made available. Personnel involved in phytomedicine processing and manufacture must be in appropriate clean and assigned working gear, including coats or approved attire, headgear or covers, nose masks, gloves, goggles, and footwear, among others. Personnel must be trained to understand the importance of personal hygiene as these directives also protect their own health and safety (Food and Drugs Authority, 2018; Rodmanee and Huang, 2013). It must be noted that not every activity will require the full complement of working gear. For instance, hand gloves may be worn where possible direct contact of hands with the product may arise. But, during packaging steps such as labelling and packing into jackets and/or cartons, hand gloves may not be required. Goggles may also be worn during activities where there is a possibility or risk of contact with the naked eye (Gouveia et al., 2015; European Commission, 2017; Rodmanee and Huang, 2013).

Figures 1 and 2 illustrate some working gear that may be used in the production setting.

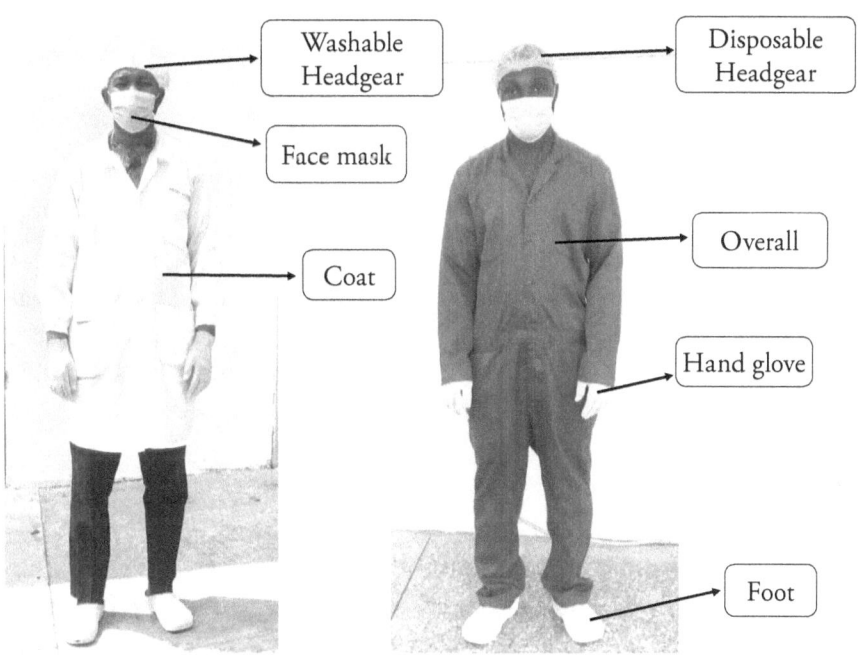

Figure 1: Protective gear: Coat Figure 2: Protective gear: Overall

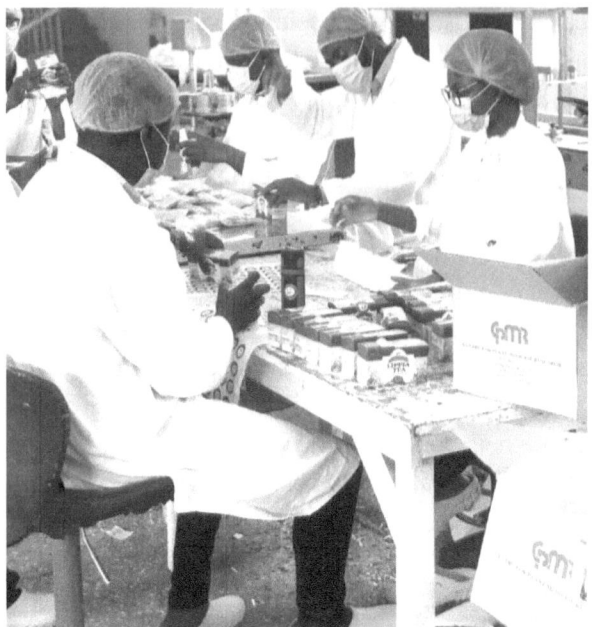

Figure 3: Small-scale phytomedicine packaging of a tea at the Centre for Plant Medicine Research (CPMR), Mampong Akuapem, Ghana

5.7 Premise

The location, design, and construction of premises should be done to suit the operations to be undertaken according to GMP. Herbal medicines, by their nature, are susceptible to microbiological contamination, degradation, and infestation by certain pests. An initial quarantine area for incoming herbal materials may help to ensure a subsequent orderly parking in storage areas (Gouveia et al., 2015; European Commission, 2017; World Health Organization, 2007).

5.7.1 Storage areas

Storage areas must be adapted and labelled for the materials to be stored, whilst ensuring that they are protected from moisture, and that insects and other animals are denied access. Storage areas for herbs, herbal preparations, and finished products should be well organised with separate areas for different materials, kept clean at all times with regularly maintenance. Spillages in storage areas should be cleaned up immediately, and reported for an assessment of the risk of contamination to other materials in storage. Materials should be stored in appropriate conditions and in a way to prevent any risk of cross-contamination (Food and Drugs Authority, 2018).

Plant materials stored in bulk should be done in aerated rooms or containers using natural or mechanical aeration and ventilation to reduce the risk of fermentation and/or mould formation. All herbal materials, whether stored in paper or fibre bags, fibre drums, or plastic bags or boxes, should be stored off the floor and adequately spaced to allow cleaning and inspection. The principle of 'first in, first out' (FIFO), must be applied to every material stored (Balekundri and Mannur, 2020; European Commission, 2017; World Health Organization, 2018).

5.7.2 Production areas

A suitable air exhaust mechanism should be used in areas where products generate dust, such as milling and drying rooms. The same precautions should be applied in areas where materials are heated or boiled to avoid the accumulation of fumes and vapours that could lead to cross-contamination (European Commission, 2017; World Health Organization, 2019).

Figure 4: Example of a phytomedicine production area, CPMR, Mampong-Akuapem, Ghana

5.7.3 Weighing areas

Weighing areas are areas where the raw plant material or milled materials are weighed. Weighing areas, whenever possible, need to be separated from the main production unit to allow easy cleaning between plant materials and herbal products. Since changes in products are much more frequent in this area, the ability to clean is essential. The design, operation, control systems, recording, and cleaning of the weighing areas must all be aimed at ensuring that there is no risk of cross-contamination. The weighing unit must be structured to control powdered products within the unit without causing contamination. All the procedures used for cleaning of the weighing area must be validated to ensure the safety and quality of the products. SOPs must be strictly followed for the cleaning method and plant materials to be used. There should be logbooks to keep a record of all weighing activities. When several plant materials are being worked on, proper identification and labelling immediately after weighing must be done to avoid mix-up or cross-contamination. There should be space for temporary storage of plant materials and easy movement of personnel (European Commission, 2017; Food and Drugs Authority, 2018; World Health Organization, 2019).

5.8 Water supply

Water supply to the manufacturing unit should be monitored and, if necessary, treated appropriately to ensure consistency of quality. There should be a written standard procedure for cleaning and prevention of contamination, which should include not just the purification system but also the distribution pipework such as the water hose. The procedure should be validated, especially the removal of disinfectant, before the system is put back into use. Appropriate water treatment plants may be used to correct defects of water from a source which may not be suitable for use in the production of phytomedicines due to lack of purity or presence of unsuitable constituents (Food and Drugs Authority, 2018; European Commission, 2017; World Health Organization, 2018).

Figure 5: Example of a demineralised (DM) water treatment plant

5.9 Qualification and validation

For small scale phytomedicine facilities, validation may apply to ensure all critical processes (including solvent purity, extraction time[s], temperature, etc.) and procedures used in manufacturing are clearly written and validated to enhance reproducibility and consistency in the quality, efficacy, and safety of the products made between batches. Changes to validated procedures may be updated to improve the quality and safety of manufactured products. The equipment used must be qualified and calibrated where applicable from time to time (Jain and Bharkatiya, 2018; Gouveia et al., 2015; World Health Organization, 2018).

5.10 Attending to complaints

Complaints normally happen in two forms: product quality complaints and adverse reaction or event complaints. A designated person or system for handling complaints must be available and clearly written. Complaints should be recorded in detail and the causes thoroughly investigated, with subsequent actions taken also recorded. Reports of adverse reactions or events should be kept in a separate register in compliance with national and international requirements. The investigation should involve comparison of the product with a reference (retained samples of the same batch) where there is a complaint about quality. For complaints of adverse reactions or events, it should be investigated to know if the reported reaction was caused by a quality problem in the product, an already reported reaction in the literature, or a new observation. Regular review of complaint records should be done to identify any specific or recurring problems requiring special attention and possible recall of marketed products. The investigation of complaints serve numerous valuable purposes. For instance, it enables producers to know if consumers are at risk and initiate any suitable action. Also, complaints help to review the product and its production processes, and institute new changes where necessary. Paying attention to the complaints of consumers and addressing them promptly, helps to create confidence in the product and the phytomedicine industry as a whole (European Commission, 2017; Nally, 2007; Patel and Chotai, 2011).

5.11 Product recalls

Recall of a product may be necessitated due to a defect realised after the product had been released on the market for consumption. The procedure for the recall of products normally depends on the national guidelines for specific countries. Small-scale phytomedicine producers should take a keen interest in procedures for the distribution of products to clients. Records of clients should be kept for tracing as much as practicable to facilitate product recalls when necessary. Instructions and written procedures for a recall activity may be given by an authorised person in the small-scale phytomedicine production department, who is responsible for the execution and coordination of recalls. Also, the personnel involved in the product recall system must follow written procedures to ensure that all defective products according to their batch numbers, are retrieved as soon as possible and stored in the appropriate segregated area. An appropriate degree of urgency must be attached to all recall activities (European Commission; 2017; Nally, 2007; World Health Organization, 2018; Food and Drugs Authority, 2018; Al-Quadeib et al., 2020)

5.12 Contract production and analyses

Contract production and analyses may involve the phytomedicine producer providing services on contract for other producers or being provided with a service for which the necessary equipment may not be available but is needed (e.g. tea bagging, extraction of essential oils, quality control analyses, etc.), at another facility. In either case, the premises and equipment used for the execution of the contract must conform to GMP standards and duly registered and certified with the appropriate regulatory body, such as the FDA in Ghana and NAFDAC in Nigeria. Owing to the high probability of production and analysis of different herbal medicinal products in the execution of such contracts, particular care must be taken to ensure validated methods are established for cleaning the equipment and premises. The contract should also specify quality and safety outcomes for the acceptance of a batch of a manufactured product (Food and Drugs Authority, 2018; Gouveia et al., 2015; World Health Organization, 2018).

5.13 Self-inspection

The aim of self-inspection is to assess the producer's compliance with GMP in all phases of production and quality control (Niazi, 2016). The self-inspection programme should be planned to identify any limitations in the execution of GMP and endorse the required curative arrangements (Chaudhari et al., 2014). This involves selecting or forming a team within the facility to check on all activities that have a bearing on the successful implementation of GMP. Making a checklist of all activities conducted by the producer and checking from time to time to ascertain if their execution is leading to expected outcomes may help in this regard. Every effort must be made to ensure that at least one member of the self-inspection team has a thorough knowledge of herbal medicines and GMP or can evaluate the implementation of GMP objectively. All recommendations for remedial action should be executed. The process for self-inspection should be acknowledged, with an active continuation programme (Food and Drugs Authority, 2018; Gouveia et al., 2015; World Health Organization, 2018).

5.14 Personnel and training

Personnel, including managers and all others who are involved in all production activities, should be well trained in relevant aspects of herbal medicine production and quality control to enhance compliance with GMP guidelines. Only authorised individuals with relevant knowledge of quality control issues should be permitted to release products. Personnel should know their individual responsibilities, which should be clearly defined and understood by the persons concerned and recorded as written descriptions. Training is an important part of GMP. It ensures that all staff and workers are informed and updated on practices that ensure the quality of all manufactured medicines. Training topics may include traditional use of herbal medicine, taxonomy and botany, pharmaceutical technology, phytochemistry, pharmacognosy, hygiene, microbiology, and related fields. Training should be done periodically, and records maintained with periodic assessments of their effectiveness (European Commission, 2017; World Health Organization, 2018; Food and Drug Authority, 2018). Training schedules should be well planned to include all staff/workers according to their specific areas of responsibility in the production of herbal medicines. Training programmes should be organised in such a way as not to affect production activities. Suggestions of practical topics that will help promote GMP in the working unit may be made by personnel of the unit to authorised personnel for consideration. A system for monitoring and evaluating the effectiveness of training programmes should be well established. All training programmes should principally be aimed at ensuring the implementation and effectiveness of the quality assurance system and GMP (Gouveia et al., 2015; Shukla, 2017; World Health Organization, 2018).

5.15 Equipment

All equipment used, such as boiling and extraction vessels, homogenizers, reservoirs, bottling lines, and tea bagging machines, must be documented and cleaned to prevent the risk of cross-contamination of products. Vacuum or wet cleaning methods are preferred for cleaning equipment. It is important to dry the equipment after wet cleaning to prevent the growth of microorganisms. The use of compressed air and brushes should be avoided as much as possible because their use increases the risk of product contamination. The equipment used for the production of herbal medicines should not contaminate or cause any adverse problem to the products. Another important aspect to inspect is the transfer of products and materials through pipelines and services supplied to the factory. It is important to clean pipelines and fold all water hoses used, attach labels, and show the contents and path of flow. Test should be conducted on the materials or products delivered through the pipelines to confirm the quality, safety, and efficacy of the products (e.g. the supply of distilled or purified water). In addition, wooden equipment should not be used unless traditionally established, and in such instances, together with other traditional equipment such as clay pots and hoppers, they should be dedicated to specific activities. They should be prevented from coming into direct contact with chemicals or contaminated material, and carefully cleaned to decrease the risk of contamination, retention of odours, and discolouration. The extent of cleaning will be based on whether the next batches are the same product or of different. It is essential that any chemicals or detergents introduced during cleaning is also completely detached to prevent contamination of the product. Wherever possible, hot water (normally >75°C) alone should be used for cleaning, and the final stains removed with cleansed water (Food and Drugs Authority, 2018; Shukla, 2017).

A confirmation programme should be based on the worst-case state, such as a relatively strong colour or odour material that is active at low levels of concentration. Also, cleaning and sanitation schedules and records of equipment cleaning should be kept for all activities. The practice of good hygiene of water systems is predominantly significant as water is such a major constituent of most products. Hot water (normally >75°C and recirculated) is a good sanitizing agent (Schmidt, 2015).

A few examples of some equipment from CPMR that may be used in the small-scale phytomedicine manufacturing industry are shown in Figures 6 to 19.

Figure 6: A liquid petroleum gas boiling vessel

Figure 7: Reservoir for decoctions

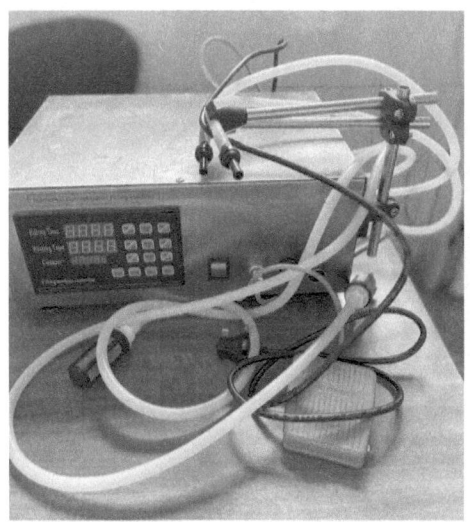

Figure 8: Mini-bottle filling machine

Figure 9: Improvised bottle filling container

Figure 10: Sifting machine

Figure 11: Tea bagging machine

Figure 12: Embosser

Figure 13: Capping machine

Figure 14: Electric stove

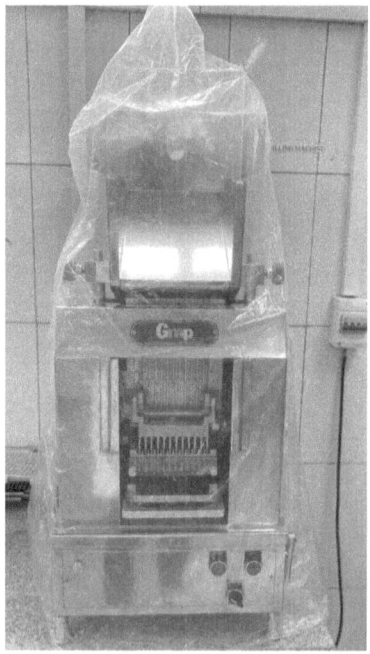

Figure 15: Capsule shell
loading machine

Figure 16: Manual capsule
filling machine

Figure 17: Oven

Figure 18: Weighing scale

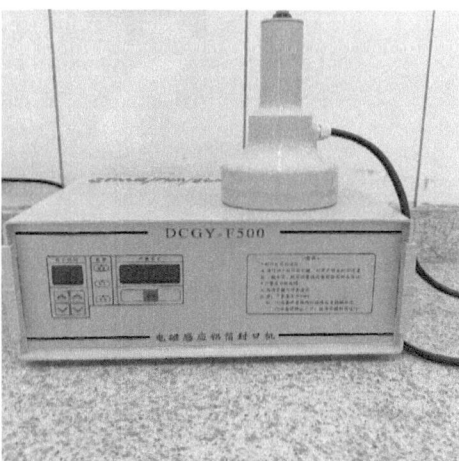

Figure 19: Heat induction sealer

5.16 Documentation

This is a very essential principle of GMP which, when applied, helps to ensure and assess the quality of products. This includes specifications.

The purpose of setting specifications is to ensure product quality and safety rather than for characterization. Specifications should exist for herbal starting materials, excipients, herbal preparations, and finished products. When dealing with starting materials, it is necessary to establish and characterise them as this goes a long way to enhance quality and ensure reproducibility.

The minimum specifications for herbal materials should as much as possible include the following:

- The family and botanical name of the plant used according to the binomial system, with the addition of the vernacular name and the therapeutic use in the country or region of origin
- The botany of the plant, such as the origin, method of cultivation, and areas it can be found or grow
- Part(s) used and in which form (e.g. whole or reduced) and the drying system where applicable
- Tests for toxic metals and for likely contaminants (microbiological, toxins, pesticide residues, foreign materials, and adulterants).

Specifications for non-herbal starting materials such as excipients and primary or printed packaging materials should be obtained from a pharmacopeial monograph where applicable (European Commission, 2017; Gouveia et al., 2015; Shukla, 2017; South Africa Health Products Regulatory Authority, 2019; World Health Organization, 2019).

5.17 Specifications for herbal preparations and finished herbal products

For herbal preparations, specifications should consist of the following:

- Clear processing instructions on the plant material, including drying, crushing, milling, and sifting methods, and processing methods for materials used in their fresh form, together with justification for that use
- The duration and temperatures required for drying and removing foreign matter. It is important to specify that drying on the ground should be avoided.
- Permissible environmental conditions, such as temperature, humidity, and standard of cleanliness, should be stated. Any treatment used to disinfect or fumigate the production unit and the method employed, should be recorded. Instructions for carrying out these procedures should be available and should include details of the process, tests, and allowable limits for residues, together with the specifications for the apparatus used. Procedures used in the blending and adjustment to obtain the phytochemical active constituents and pharmacological activity should be documented. Measures used in discarding expired and spent herbal products should also be recorded (European Commission, 2017; Food and Drugs Authority, 2018; World Health Organization, 2019).

For small-scale phytomedicine producers, the main focus should be on ensuring that products have been prepared with the minimum acceptable limits of contaminants. Therefore, tests for microbiological contamination should be paramount, and producers should be encouraged to acquire test kits, which may be able to help determine the presence of harmful microbial agents, particularly *Escherichia coli*, *Staphylococcus aureus,* yeast and moulds. Other tests that can be simply done without relatively expensive equipment, include assessment of physical appearance, such as colour, odour, taste, texture, and size. Tests for pH and uniformity of weight and disintegration tests for tablets, single-dose powders, suppositories, capsules, and herbal tea in sachets may also be done where applicable (European Commission, 2017; World Health Organization, 2019). Other tests such as those for toxicants and chromatographic fingerprinting of relevant constituents or markers, may be done periodically as it may not be practical to do them on every batch of a particular product.

5.18 Good practices in phytomedicine production

In the production of phytomedicines, the production processes and methods should be clearly specified to ensure the quality, safety, and efficacy of the products. The cultivation and harvesting of raw plant material for the production of herbal medicine should follow SOPs and *Guidelines on Good Agriculture and Collection Practices (GACP) for Medicinal Plants*. Processing of the raw plant materials such as cutting, sorting, washing, and drying should be guided by GMP.

The production of powdered herbs or teas should be guided by GMP. Finished herbal products manufactured through fermentation should also follow GMP for all the production stages to ensure quality, safety, and efficacy of the product (European Commission, 2017; Food and Drugs Authority, 2018; Kwesiga, 2021; World Health Organization, 2019).

5.19 General practices to consider in phytomedicine production

Plant materials harvested from the farm should be unloaded and arranged on pallets. They should not have direct contact with soil or the ground after harvesting. They must be handled so that they are not exposed to rain and microbial contamination. The drying of harvested plant materials should be based on specification or as needed.

The production area for preparation of phytomedicines and the environment should be cleaned routinely. SOPs should be developed to cover operations involving equipment, personal hygiene, production floor, and washrooms and eliminate microbial contamination. Cleaning should be scheduled for at least twice a month.

Appropriate cleaning methods should be chosen for the cleaning of herbal materials. When cleaning dried herbal materials, washing them should be avoided; instead, an air duster should be used.

Harvested plant materials and processed materials should be stored in separate storage rooms and well labelled with their batch numbers and dates of processing to avoid cross-contamination.

Production personnel should adhere to safety rules and SOPs to avoid contamination during production of phytomedicines.

The production of phytomedicines should take place in accordance with the master process documents. The necessary procedures should be used and recorded during the production process. Records should be kept on production inputs, including plant materials, packaging materials, and equipment. SOPs should be followed to ensure that there is no mix-up during packing and labelling of products.

Smoking, eating, and drinking should not be allowed in the manufacturing area.

Personal hygiene procedures, including wearing protective clothing, apply to all people entering into production areas.

Direct contact between the operator and the product should be avoided wherever possible. If direct handling is unavoidable, then gloves should be worn, and if appropriate, they should be disinfected after being worn.

Repair and maintenance activities should be carried out in a way that does not present any risk to the product or personnel. Maintenance of equipment should not be done during production hours. Instead, time should be allocated for repair and maintenance of equipment. Any emergency work in a working area should be followed by a thorough clean up and disinfection of the area before manufacturing resumes, and an area clearance check by the production manager should be done after a maintenance activity (Food and Drugs Authority, 2018; Gouveia et al., 2015; Shukla, 2017; World Health Organization, 2019).

5.20 Mixing of batches and blending

The mixing of batches of herbal plant materials ensures that the beneficial constituents are generally uniform. It is very important to document and specify any substance added for the purposes of uniformity. Blending of different batches of known constituents could be accepted. The blending process should be effectively stated and documented to ensure easy traceability. Batches of herbal medicine that are blended together should be specified and have an established process. Individual testing of the batches should be done prior to blending, and all blended batches should follow the expiry date of the previous batch (European Commission, 2017; Yan and Qu, 2013; World Health Organization, 2019).

5.21 Good practices in quality control

Quality control is the part of GMP that is concerned with sampling, specifications, testing, organisation, documentation, and release procedures using validated methods implemented by trained and experienced personnel. Making good decisions pertaining to the quality control process, not only depending on the laboratory operation, should be emphasised to ensure the quality of the herbal products. Quality control staff must therefore sign off on all manufacturing procedures that are relevant to product quality (Food and Drugs Authority, 2018; Gouveia et al., 2015).

5.21.1 General

Personnel with qualifications or knowledge in herbal medicine quality control should perform microbial tests, detect adulteration, and identify a lack of standardisation in the herbal medicine. The quality of herbal preparations and finished herbal products should be controlled and monitored by the quality control personnel (World Health Organization, 2019).

5.21.2 Sampling

Herbal materials have an element of heterogeneity; they are collections of individual plants and need to be sampled. Sampling should be approved by qualified personnel. The most essential aspect of it is that it is symbolic of the batch and is in line with SOPs (European Commission, 2017; World Health Organization, 2018).

5.21.3 Testing

The quality, safety, and efficacy of herbal materials and finished herbal products should be tested in accordance with written methods using equipment that is calibrated regularly. The herbal preparation and finished herbal products should be subjected to microbial, phytochemical constituent and pharmacological analysis. For example, active compounds such as flavonoids, saponins, and reducing sugars could be analysed. Physical identification of the herbal products should be done, if applicable, to identify the active ingredients, and at least the main ingredients should be indicated on the label. Retained samples of herbal materials and finished products should be made available for use in retesting or virtual and consistent testing. For instance, organoleptic properties and visual examination can be done. The type of test to be conducted on herbal preparations and finished herbal products should be selected based on the specification to ensure quality, safety, and efficacy. The test of the herbal products must be repeatable over a number of equal tests. Every batch of herbal products must be tested (European Commission, 2017; Food and Drugs Authority, 2018; Gouveia et al., 2015; Shukla, 2017, World Health Organization, 2019).

5.21.4 Stability studies

The stability of herbal products can be defined as the time during which the herbal products preserve their microbial, physical, chemical, and pharmacokinetic effects and quality for the duration of their shelf life and time of production. 'Shelf life' is a practical term used to signify the stability of the product and can be defined as a reduction in the concentration of the product of up to 90% of its initial concentration. It is also known as the expiry date. The expiry date for a herbal product should be given to assist stability studies on the product under specified storage conditions. The shelf life of the finished herbal product is predicted when there is stability data. Stability studies of herbal preparations and finished herbal products help in the development of new drugs and formulations (Saranjit, 2006), and also ensure that product quality, safety, and efficacy are maintained throughout the shelf life of the herbal products (Aashigari et al., 2018; Kumadoh et al., 2020). Stability studies on finished herbal products ensures the right medication and dosage form of herbal medicine. It also helps in preventing toxicity resulting from decomposition of the finished product. Transporting the finished herbal products for marketing or from one place to another could lead to changes in their physical and chemical properties (Thorat et al., 2014). During stability studies, the preservatives and stabilizers should be observed. Three batches of finished herbal products meant for marketing, should be selected and used in the stability investigation programme to check the expiry date. But when stability data from preceding studies reveals that selected batches of the products are expected to stay stable for at least two years, less than three batches can be used. Regular stability studies should be undertaken, which would normally encompass one batch of herbal product being used in the stability investigation programme each year (Kaur and Gulshan, 2020; World Health Organization, 2018).

5.22 Packaging materials and labelling

Packaging materials such as cartons, bottles, and labels should be stored well. Bottles should be well arranged on pallets with labels packed in cabinets, and washed thoroughly before being used for bottling the herbal medicine. All labels issued for use should be controlled and documented. Labels should bear the product name, indication, dose and dosage form, active ingredients, adverse reactions, cautions, batch number, manufacturing date, expiry date, and FDA number (European Commission, 2017; Food and Drugs Authority, 2018; Gouveia et al., 2015; Shukla, 2017; World Health Organization, 2018).

Some examples are shown in Annex 2: Table 1: General Information for Labels of Phytomedicine Products and Table 2: General Information for Labels of Some Phytomedicine Categories (Decoction, Teas, Powders, Ointments, Capsules, and Tablets).

5.23 Conclusion

Small-scale phytomedicine production must strictly comply with GMP procedures, to ensure quality, safety, and consistency in the products given to patients. Applying the GMP concepts will help in achieving the definitive goals of protection of the health and safety of the patient, as well as producing good-quality medicines. Quality can be accomplished only through careful planning and practical implementation of GMP.

References

1. Aashigari S, Goud R, Sneha S, Vykuntam U, Potnuri NR. (2018). Stability studies of pharmaceutical products. World Journal of Pharmaceutical Resience, *8*, 479–492.
2. Abubakar AR, Haque M (2020). Preparation of Medicinal Plants: Basic Extraction and Fractionation Procedures for Experimental Purposes. Journal of pharmacy & bioallied sciences, 12(1):1–10.
3. Al Ph, Hüseyin Ar, Ma R, et al. (2012). Standard operating procedures (SOPs) for the development of Unani polyherbal formulation "habb-e-azaraqi" and its physicochemical analysis. International Journal of Institutional Pharmacy and Life Sciences 2(4), 2249–6807.

4. Alison H, Beverley DG (2011). Pharmaceutical excipients – where do we begin? Aust Prescr, 34, 112–114, https://doi.org/10.18773/austprescr.2011.060

5. AlQuadeib BT, Alfagih IM, Alnahdi AH et al., (2020). Medicine recalls in Saudi Arabia: a restrospective review of drug alerts (January 2010-January 2019), Future Journal of Pharmaceutical Science., 6, 91, https://doi.org/10.1186/s43094-020-00112-3

6. Osei-Asare C, Owusu F, Entsie P et al. (2021). Formulation and In Vitro Evaluation of Oral Capsules from Liquid Herbal Antimalarials Marketed in Ghana. Journal of tropical medicine, 2021, 6694664. https://doi.org/10.1155/2021/6694664

7. Balekundri A, Mannur V (2020). Quality control of the traditional herbs and herbal products: a review. Future Journal of Pharmaceutical Science 6, 67. https://doi.org/10.1186/s43094-020-00091-5

8. Benzie F, Watchtel-Galor S (2011). Herbal Medicine, Biomolecular and Clinical Aspects. 2nd Ed., London. CRC Press.

9. Boadu AA, Asase A (2017). Documentation of Herbal Medicines Used for the Treatment and Management of Human Diseases by Some Communities in Southern Ghana. Evidence-based complementary and alternative medicine; *eCAM, 2017*, 3043061. https://doi.org/10.1155/2017/3043061.

10. Centre for Plant Medicine Research. 2022 (A). Standard Operating Procedures for Production Department. CPMR, Ghana.

11. Centre for Plant Medicine Research. 2022 (B). Labels for *Asmodium* and *Lippia* tea Jackets. Production Department. CPMR, Ghana.

12. Chaudhari VK, Yadav V, Verma PK, Singh AK (2014). A Review on Good Manufacturing Practice (GMP) for Medicinal Products; PharmaTutor; 2(9); 8–19.

13. Djordjevic SM (2017). 'From Medicinal Plant Raw Material to Herbal Remedies', in H. A. El-Shemy (ed.), Aromatic and Medicinal Plants - Back to Nature, Intech Open, London. 10.5772/66618

14. European Commission (2017). Good Manufacturing Practice. Guidelines on Good Manufacturing Practice specific to Advanced

Therapy Medicinal Products, 4, European Commission. Eudralex-EU, Brussels

15. Food and Drugs Authority (2018). Guidelines For Conducting Cgmp Inspection Of Herbal Manufacturing Facilities Located In Ghana; Fda/Drid/Did/Gl-Gmp-H/2018/01

16. Gouveia BG, Rijo P, Tânia SG, Catarina PR (2015). Good manufacturing practices for medicinal products for human use. Journal of Pharmacy & Bioallied Sciences. 7, 87–96. 10.4103/0975-7406.154424.

17. Guo CA, Ding XY, Addi YW *et al.,* (2022). An ethnobotany survey of wild plants used by the Tibetan people of the Yadong River Valley, Tibet, China. Journal Ethnobiology Ethnomedicine 18, 28. https://doi.org/10.1186/s13002-022-00518-8

18. Jain K, Bharkatiya M (2018). Qualification of Equipment: A systematic approach. International Journal of Pharmaceutical Sciences. (8), 7–14.

19. Kaur J, Gulshan B (2020). WHO prescribed shelf-life assessment of Syzygium cumini extract through chromatographic and biological activity analyses, Journal of Ayurveda and Integrative Medicine, 11(3), 294–300.

20. Kumadoh D, Ofori-Kwakye K (2017). Dosage Forms of Herbal Medicinal Products and their Stability Considerations-An Overview, Journal of Critical Reviews, 4 (4): 1–8.

21. Kumadoh D, Kwakye KO, Kuntworbe N, Adi-Dako O, Appenahier JA (2020). Determination of shelf life of four herbal medicinal products using high-performance liquid chromatography analyses of markers and the Systat Sigmaplot software. Journal of Applied Pharmaceutical Science 10(06):072–080.

22. Kwesiga V, Ekeocha Z, Byrn S, Clase K. (2021). "Compliance to GMP guidelines for Herbal Manufacturers in East Africa: A Position Paper" (2021). BIRS Africa Technical Reports. Paper 7. http://dx.doi.org/10.5703/1288284317428

23. Martindale: The complete drug reference. 37th ed. London: Pharmaceutical Press; 2011. (Electronic and hard copy available)

24. Nally JD (2007). Good manufacturing practices for pharmaceuticals, 6th Edition, Informa healthcare USA, Inc., ISBN 10:0-8593-3972-3 & ISBN13, 978-0-8493-3972- 1.

25. Natako L (2006). "Honouring the African Traditional Herbalist" African Traditional Herbal Research Clinic Newsletter. Special Edition—HIV/AIDS.25 years.1 (10)

26. Ndhlala AR, Stafford G.I, Finnie JF, Van Staden J (2011). Commercial herbal preparations in KwaZulu-Natal, South Africa: The urban face of traditional medicine. South African Journal of Botany.77,830–843. 10.1016/j.sajb.2011.09.002.

27. Niazi SK (2016). Handbook of Pharmaceutical Manufacturing Formulations: Over the Counter Products,480

28. Nishteswar K, Tavhare S (2014). Collection Practices of Medicinal Plants - Vedic, Ayurvedic and Modern Perspectives. International Journal of Pharmaceutical and Biological Archives. (5). 54–61.

29. Ozioma EJ, Nwamaka Chinwe OA (2019). 'Herbal Medicines in African Traditional Medicine', in P. F. Builders (ed.), Herbal Medicine, IntechOpen, London. 10.5772/intechopen.80348.

30. Pandey AK, Savita (2017). Harvesting and post-harvest processing of medicinal plants: Problems and prospects. The Pharma Innovation Journal; 6(12), 229–235

31. Patel KT, Chotai, NP (2011). Documentation and record, Harmonized GMP requirement. Journal of young pharmacist, 3, 138–150

32. Porwal O, Singh S, Patel D et al., (2020). Chapter 2; Cultivation, Collection and Processing of Medicinal Plants. In Bioactive Phytochemicals: Drug Discovery to Product Development 1–16. 10.2174/9789811464485120010005

33. Prakash G, Abinash CS, Pandey S (2020). Guidelines on Stability Studies of Pharmaceutical Products and Shelf-Life Estimation, International Journal of Advances in Pharmacy and Biotechnology, 06(01), 15–23

34. Rodmanee S, Huang W (2013). Hygiene and Manufacturing Practices, Interagency Collaboration, and a Proposal for Improvement: A Case Study of Community Food Enterprise in Thailand. International journal of social science and humanity, 222–226.

35. Runde FD, Conor MS, Janina, S (2021). Supporting Small and Medium Enterprises in Sub-Saharan Africa through Blended Finance, CSIS (Centre for Strategic and International Studies) Briefs, 1–8.

36. Schippmann U, Leaman D, Cunningham AB (2006). Comparison Of Cultivation and Wild Collection of Medicinal and Aromatic Plants under Sustainability Aspects. In R.J. Bogers, L.E. Craker and D. Lange (eds.), Medicinal and Aromatic Plants, 75–95., Springer. Printed in the Netherland

37. Schmidt RH (2015). Basic Elements of Equipment Cleaning and Sanitizing in Food Processing and Handling Operations 1. U.S. Department of Agriculture, University of Florida, IFAS, Florida.

38. Shukla J (2017). Good manufacturing Practice: An Overview. Available at https://www.researchgate.net/publication/320373559_Good_manufacturing_Practice

39. South Africa Health Products Regulatory Authority (2019). SA Guide to Good Manufacturing Practice, 2019. Jul 19, 97

40. Thorat P, Warad S, Solunke R, et al., (2014). Stability Study of Dosage Form: An Innovative Step. World Journal of Pharmacy and Pharmaceutical Sciences, 3(2), 1031–1050

41. United States Food and Drug (2020). Facts About the Current Good Manufacturing Practices (CGMPs). https://www.fda.gov/ drugs/pharmaceutical-quality-resources/ facts-about-current-good-manufacturing-practices-cgmps

42. World Health organization (2007). WHO guidelines on good manufacturing

43. World Health organization (2018). WHO Expert Committee on Specifications for Pharmaceutical Preparations Fifty-second report. WHO Technical Report Series, No. 1010, 2018

44. World Health organization (2019). WHO global report on traditional and complementary medicine 2019 ISBN 978-92-4-151543-6 World Health Organization 2019, Printed in Luxembourg. https://www.who.int/teams/health-product-and-policy-standards/standards-and-specifications/gmp

45. Yan B, Qu H (2013). An approach to optimize the batch mixing process for improving the quality consistency of the products

made from traditional Chinese medicines. Journal of Zhejiang University. Science. B. 14. 1041–1048.

ANNEX 1

- **Sample Standard Operating Procedure (Centre for Plant Medicine Research, 2022A)**

Standard Operating Procedure for Personal Hygiene and Discipline

Written by	Signature	Date	Approved by	Signature	Date

Circulation

This procedure is issued on a controlled basis to:

1. Head of production	2. Production technologist
3. All staff	4.

1.0 Purpose: To ensure good hygiene practices and personnel discipline on production floors and stores

2.0 Scope: Production staff and visitors

Who: Head of production/production technologist, all staff

What: Procedure for personal hygiene and discipline

Where: To be observed in the production department, production floor

3.0 References: WHO Guidelines for Good Manufacturing Practices for Herbal Medicines

4.0 Definition: None

5.0 Procedure

5.1 Wear nose masks and hand gloves at all times during production.

5.2 Eating, smoking, chewing, drinking, and keeping foods, drinks, and personal materials are not permitted on the production floor and storage areas or in any other areas where they might adversely influence product quality.

5.3 Excessive noise making and laughing is not permitted on the production floor.

5.4 Avoid direct contact between the operator's hands and starting materials, primary packaging materials, and intermediate or bulk products.

5.5 Report to your immediate supervisor any conditions (relating to plant, equipment, or personnel) that may adversely affect the products.

5.6 Observe high level of personal hygiene during manufacturing processes. Wash hands before entering production areas. Observe pasted signs and instructions.

5.7 Wear clean body coverings appropriate to your duties, including appropriate coats, footwear, and hair covering.

6.0 Related documents: None

ANNEX 2

Table 1: General Information for Labels of Phytomedicine Products (Centre for Plant Medicine Research, 2022B)

Term	Definition
Indication	A medical condition that a product is to be used to treat. For example: herbal medicine for supporting the immune system
Dosage	How a herbal product should be taken to realise expected outcomes. For example: Adults: Two tablespoonful (30 mL) three times daily after meals Children: Twelve (12) years and above one tablespoonful (15 mL) three times daily after meals
Active ingredient	The substance(s) that has/have the desired therapeutic effect For example: *Desmodium adscenens, Lippia multiflora*
Adverse reaction	An unwanted effect caused by the administration of a drug. If none has been reported, it can be represented as: None reported
Caution	Care taken to avoid danger or risk of the ingested herbal product. For example: Not recommended for pregnant women, nursing mothers, and children below twelve years. Keep out of reach of children.
FDA registration number	A number denoting that the product has been registered by the FDA. For example: FDA/HD119-10357
Batch number	A number denoting the batch of the herbal product For example: AS 22001
Manufacturing date	A date a herbal product is produced For example: 8 February 2022
Expiry date	A date after which a herbal product is not suitable for use For example: 8 February 2024

Volume	The hypothetical fluid *volume* through which the herbal product is dispersed. For example: 330 mL

Table 2: General Information for Labels of Some Phytomedicine Categories (Decoction, Teas, Powders, Ointments, Capsules, and Tablets)

Information	Example(s)
Directions	Example (A) Teas/Infusions: Place two bags into two teacups of freshly boiled hot water (approximately 500 mL) and infuse for five minutes. Example (B) Powder/Decoctions: Put four teaspoonful of herbal material into one cup (300 mL) of boiling water, and allow to boil for three minutes. Strain and decant. Or add one teaspoonful of powder to any beverage, porridge, or soup. Example (C) Roots/Tinctures: Put half content (amount 40 gm) into 600 mL of gin and allow to stand for five to seven days. Example (D) Ointment/Cream: Apply to the affected parts two times daily. Example (E) Capsules/Tablets
Dosage	Example (A): Drink one prepared cup of tea two times daily or take one teaspoonful two times daily for powders. Example (A) Adults: Two tablespoonful (30 mL) three times daily after meals Example (A) Children: Twelve (12) years and above one tablespoonful (15mL) three times daily after meals for decoctions. Example (B): Add 5 mL to 40 mL of water, mix and drink all for tincture.

	Example (D): Apply the ointment to the affected area twice daily.
	Example (E): Take one tot glass three times daily or chew three to six pieces of roots daily.
	Example (F) Take two or three capsules/tablets three times daily.
Active ingredient	The active ingredients/plant materials depend on the herbal product to be prepared, for example: *Desmodium adscendens, Bridelia ferruginea, Capparis erythrocarpus, etc.*
	Alcohol percent in the tinctures must be detected and indicated on the labels, e.g. 10% v/v or 25%v/v
Adverse reaction	None reported
Caution	Not recommended for pregnant women, nursing mothers, and children below 12 years
	Keep out of reach of children.
FDA number	FDA/HD119-10357
Batch number	BT 22001
Manufacturing date	8 February 2022
Expiry date	8 February 2024
Weight	40 g or 40 gm or appropriate weight per unit for powders, teas, ointments, roots, capsules, tablets, etc.
Volume	330 mL or 120 mL or appropriate volume for decoction, tinctures, etc., depending on the bottle used

6

Understanding and Appreciating Sector Potentials – Investment, Policymaking, Stakeholder Engagement

Nanlop Adenike Ogbureke

Office of the Director General, West African Health Organization, Bobo Dioulasso, Burkina Faso

6.1 Introduction

The debate on traditional medicine (TM) has been an ongoing one and is mostly hinged on theories and beliefs with little or no evidence on the negative impacts (WHO, 2004). However, over the years, the world and indeed the West Africa region have seen a significant shift in ideologies and mindsets regarding the benefits of the use of traditional medicines as well as the efficacy and effectiveness in the prevention, management, and treatment of illnesses/diseases (Oyebode et al., 2016). It is necessary to define health to begin to expound on and appreciate the place of TM, the socioeconomic and cultural benefits, as well as its role in contributing to the attainment of improved health outcomes not only in the West Africa region, but also globally.

According to the World Health Organization (WHO), health is a state of complete physical, mental, and social well-being and not merely the absence of disease or infirmity. Other definitions of health indicate how the thinking has evolved from a 'medical model', which views health through a treatment lens focused on physical illnesses that impact on the human body, to a 'holistic model', which includes other components of mental and social well-being – an apt example is the WHO definition of health of 1947. Subsequently, there is now a shift to a more integrated wellness model championed by WHO as part of its WHO health promotion initiative and amplified during the Ottawa Charter of 1986 (WHO EURO, Ottawa Charter for Health Promotion,1986).

Other related definitions view health in terms of the capability of people or groups of people to successfully cope in the face of significant adversity or risks. It is justifiable to state that the definition of traditional medicine as collection of knowledge, skills, and practices for the maintenance of health, as well as prevention, diagnosis, improvement, or treatment of physical and mental diseases (Benzie and Wachtel-Galor, 2011; Smith et al., 2014) brings it all together. Of much more pertinence here is the definition that can be best described as an ecological definition that brings together human, animal, and environment within the context of resilience, shifting economic realities and/or natural disasters.

In considering these definitions, there is the emergence of the concept of medicine as a science and practice of caring for persons with ill health. As Pan et al. (2014) discussed in the paper 'Historical Perspective of Traditional Indigenous Medical Practices', this dates to the prehistoric times where medicine was practiced as having connections to religious and philosophical beliefs of the local culture of the time. This was mainly seen in the use of herbs alongside prayers for health care, giving rise to the term 'medicine man' or 'herbalist'.

In recent centuries and since the advent of modern science, most medicine has become a combination of art and science (both basic and applied, under the umbrella of medical science). Pre-scientific forms of medicine are now known as traditional medicine or folk medicine, which remains commonly used in the absence of scientific medicine and are thus called alternative medicine (Fokunang, 2011). Alternative treatments outside of scientific medicine having safety and efficacy concerns can be described as quackery. The perceptions and beliefs that drive alternate or traditional medicine options as sub-standard and not meeting ethical standards of medical and pharmaceutical practices portend a challenge for the advancement of phytomedicines globally and particularly in Africa, even with evidence of the efficacy of these practices and science. According to Abdullahi (2011), the unaddressed global impact of European colonisation on TM and traditional medicine practice (TMP) as documented and discussed has been devalued, subjugated, co-opted, and in some cases decimated.

This is largely attributed to the argument that medicine is all-encompassing of evolving healthcare practices with the primary objective of maintaining and restoring health through the prevention and treatment of illness. This strengthens the pre-supposition that challenges the efficacy and effectiveness of traditional medicines in the prevention, management, and treatment of illness as well as in improving the overall health and well-being of individuals. Traditional medicine, which existed in human societies before the application of modern science to health, varies widely in keeping with the societal and cultural heritage of different countries. Every human community responds to the challenge of maintaining health and treating diseases by developing a medical system. Thus, traditional medicine has been practiced to some degree in all cultures and should, therefore, be prioritised particularly in contexts where access to and affordability of quality healthcare services is slow paced and fragility and emerging and reoccurring epidemics contribute to impeding and further weakening existing health systems.

6.2 Building an investment case for traditional medicine

An understanding of contemporary medicine is hinged on the application of multiple disciplines such as genetics, biomedical sciences, and research and medical technology to diagnose, treat, and prevent injury and disease, as well as the development of pharmaceutical products, psychotherapy, and surgery. While modern medicine is often described as the preferred form of medicine, traditional medicine as the ancient and culture-bound medical practice is developing rapidly, with significant contributions to disease control, prevention, and treatment as well as to therapeutic care. Amidst this lies the relationship between both practices and presents emerging opportunities that intrinsically links both to the main objective of ensuring improved health outcomes of populations.

Therefore, with over 80% of the population using traditional remedies rather than modern medicine for primary health care, the space for thriving traditional medicine practices is seen to be expanding across various climes, with Asia and Africa profiting from advanced utilisation and commercialisation (Oyebode, 2016). Interestingly, even as modern medicine continues to improve in terms of knowledge and use of modern technologies, the rapidly increasing interest in and utilisation of traditional medicine presents a huge and evolving market because of the growing number of people using it for different ailments and other diseases. This is also recognising the inadequacy of access to and affordability of allopathic medicines and Western forms of treatments and modern medical care by the majority of people in Africa, particularly for some ailments such as malaria and/or HIV/ AIDS (Sato, 2012).

The demand for herbal products and medicines by the public presents an opportunity to promote collaboration for its advancement through research, regulation, and certification by relevant regulatory bodies (Tilburt and Kaptchuk, 2008) such as the WHO International Regulatory Cooperation for Herbal Medicines (IRCH), established in 2006 to serve as a global network of regulatory authorities responsible for the regulation of herbal medicines to protect and promote public health and safety. This is even more relevant at a time where there are reoccurring and new epidemics, continued uncontrolled use, and overarching possibilities of integrating both traditional and modern medicine.

The positioning of traditional medicine can be further argued and instituted as stated in the WHO TM Strategy (2014–2023), developed in response to the World Health Assembly resolution on traditional medicine (WHOA62.13), where it is highlighted as an important and often underestimated and overlooked part of health services. The pharmacopeia of the West African Health Organization (WAHO) reiterates the benefits of TM, its use within the region, and its potential contribution toward improved health outcomes on the regional population.

While the focus of public health may be defining the risks and benefits of herbal medicines already in use, entrepreneurs and corporations hope herbal medicines may yield immediate returns from herbal medicine sales or yield clues to promising chemical compounds for future pharmaceutical development. They test and analyse individual herbs and their components in state-of-the-art high-throughput screening systems, hoping to isolate therapeutic phytochemicals or biologically active functional components. In an article written by Richard Stone and Hao Xin in 2006 and published in the online news journal, *Science*, Novartis was reported to have made a commitment to invest over US$100 million to investigate traditional medicine in Shanghai alone. Harnessing the opportunities of commercialised TM and the huge market is evident in countries like China, where in 2005, traditional medicines worth US$14 billion were sold, and Brazil, which saw revenues of US$160 million from traditional therapies. According to the SciDevNet article 'Integrating Modern and Traditional Medicine: Facts and Figures (2010)', both examples are part of a growing global market of more than US$60 billion, which does not include the total out-of-pocket spending of US$14.8 billion for natural products in the United States in 2008 alone. While it is difficult to ascertain the total market expenditure on TM in West Africa, these figures present a strong investment case for traditional medicine.

6.3 Traditional medicine and the policy debate

The need to explore these alternatives demonstrates the importance of putting in place regulations, consolidating best practices, and exploring the socioeconomic benefits and opportunities. This is even more critical when viewed against the number of people around the world who have treated the sick with herbal or animal-derived remedies handed down through generations and the antecedent weak primary healthcare systems across the West Africa region as well as the cost of quality healthcare services that impacts heavily on the large poor and vulnerable populations. There is a considerably large proportion of people depending on traditional remedies, medicines, and practices for solutions to health issues. The place of research is also important to generate additional evidence of the effectiveness and safety of traditional medicines and drive demand, galvanise political commitment, and increase stakeholder investments. Furthermore, interrogating the science behind traditional medicine will provide the data needed for framing a scientific approach to policy implementation, engagement, and regulation and subsequently integrating TM products, practitioners, and practice into health systems as appropriate (Gyasi et al., 2017).

It is against this backdrop that the policy question emerges as a crucial point not only for sustainability but also for institutionalisation, practice, and synergy between both strands of traditional and modern medicine. The use of evidence has already been highlighted in some sections of the book; however, the elephant in the room remains. The consideration of TM by policymakers within the framework of national, regional, and global governments should be upheld through the intentional act of creating laws or setting standards for practice.

The 2014–2023 TM strategy was developed first to provide member states with the guidelines required to harness these potential contributions to people-centred health and wellness and second to promote safe and effective use of TM through the regulation, research, and integration of products, practices, and practitioners into the health system. As part of the efforts towards achieving the universality of TM, its implementation, and importantly its acceptability and integration within the mainstream health system, existing models and best practices across member states should be accessed, recorded, recommended, and adapted for use. Furthermore, and hinged on the 2009 Executive Board and World Health Assembly Resolutions on Traditional Medicine, the WHO TM strategy and the WAHO Pharmacopeia (2013, 2020) present policymakers with valuable evidence and a roadmap for evidence-informed policy frameworks as a key component of the universal health coverage (UHC) agenda generally and primary health care (PHC) specifically.

6.4 Integrating TM into health systems and curricula

Strong primary health care (PHC) is recognised as integral to achieving UHC (van Weel and Kidd, 2018). This is even particularly so because of the positioning of primary-level care as essential and ideally affordable and accessible to mostly poor and often vulnerable and rural populations. According to the WHO Primary Health Care Factsheet accessed in April 2021, PHC is rooted in a commitment to social justice, equity, solidarity, and participation based on the recognition that health is one of the fundamental rights of every human being. The local community is central to the operations of primary health care as it relies on access to skilled health providers and professionals with the mandate to address the health problems of the people.

As more and more people use TM and herbal options, the more imperative it becomes to ensure the integration of TM into health systems, particularly the primary healthcare system. The pertinence of TM in improving patient experience and population health should be made clear to policymakers, policy influencers, and other key stakeholders such as pharmaceutical manufacturers and healthcare consumers. While this is necessary for action, academic institutions and their role in preparing health professionals with expertise in TM practice and research should be emphasised with pragmatic solutions that align to the local socioeconomic, cultural, and political situation. It is also important to bear in mind the dependency of the largest proportion of the population in rural areas across the region, whose primary health providers are native healers.

Considering the ratio of traditional healers to the population in Africa is 1:500 as compared to the ratio of medical doctors to the population of 1:40,000, it is commendable that that some member countries have included TM as part of their university curricula for health profession students and a few other countries have established national research institutes (Abdullahi, 2011). In addition, it is worth noting that herbal medicines are used as the first line of treatment for 60% of children presenting with fever in countries such as Ghana, Mali, Zambia, and Nigeria (WHO, 2002).

In view of these laudable achievements, one of the key recommendations that should be explored is the establishment of centres of excellence for TM to create operational initiatives aimed at increasing the productivity of affordable quality phytomedicines and the establishment of a regional community of practice, led by the West African Health Organization (WAHO), not only as the health institution of the ECOWAS region but also as the primary hub of traditional medicine research and promotion and through the WAHO technical advisory Committee for TM set up in 2021.

Establishing a community of practice will contribute to bridging the gap between TM and modern medicine practitioners as it will encourage knowledge sharing and give members a networking platform for knowledge, information, and experience sharing and turning knowledge and research into practice. The combination of practitioner knowledge and experience will be invaluable support for evidence-based practice and will further strengthen the argument for the integration of TM into health systems and as part of formal health and medical academic curricula.

6.5 The place of evidence-based research for policy formulation and implementation

According to the 2019 WHO Global Report on Traditional and Complementary Medicine, there is an upward trend of about 87% of the number of member states who formally acknowledge the use of TM. Out of this proportion, 100 member states have a national policy on TM, and 124 member states have a national regulation of herbal medicines. In the West Africa sub-region, this includes Ghana, Nigeria, The Gambia, and Senegal.

It can be noted that while several countries have national TM policies, strategic documents, and in some cases, legal instruments, which largely align with the WHO TM strategy, set the stage required to drive implementation. In the region, the social value of traditional medicine research should be advocated for and implemented alongside public health officials. This will strengthen collaboration between both TM and modern medicine, particularly with regard to defining the safety, effectiveness, efficacy, and efficiency of herbal medicines for the treatment and management of diseases and other health conditions such as malaria while mitigating potential harm that may arise from illegal use of certain herbs.

Accordingly, in public health bodies in many countries, there has been increasingly more financial resources allocated for herbal medicine research. To put it in perspective, the journal *Frontiers in Pharmacology* published an article in April 2022 entitled 'Safety of Using Traditional Chinese Medicine Injections in Primary Medical Institutions: Based on the Spontaneous Reporting Systems 2016–2020 in Henan Province, China'. The article provides the example of China's recently launched safety research programme focusing on herbal medicine injections from traditional Chinese medicine (Yan et al., 2022). Furthermore, South Africa added traditional medicines investigation within its national drug policy (WHO/AFRO, 2004). The National Cancer Institute of the United States committed funds to study a range of traditional therapies. While this scale of investment pales in comparison to the total research and development expenses of the pharmaceutical industry, nevertheless, it reflects genuine public, industry, and governmental interest in this area.

Though the public good of TM has received more visibility and priority, there is still a lot to be done especially as entrepreneurs and corporations hope herbal medicines may yield immediate returns from sales or yield clues to promising chemical compounds for future pharmaceutical development. This is in view of the concern of public health entities regarding the importance of defining the risks and benefits of herbal medicines already in use. Irrespective of the scepticisms, individual herbs, or their components, are being analysed and tested in state-of-the-art high-throughput screening systems, hoping to isolate therapeutic phytochemicals or biologically active functional components. A relevant example is the reported investment of over US$100 million by Novartis in 2006, to investigate traditional medicine in Shanghai alone.

Nongovernmental organisations such as the Association for the Promotion of Traditional Medicine (PROMETRA), based in Dakar, Senegal, are dedicated to preserving and restoring African traditional medicine and indigenous science, being primarily interested in preserving indigenous medical knowledge and improving population health. In some cases, governments, such as in Nigeria, whose president established a national committee on traditional medicine with the expressed desire to boost Nigeria's market share, may want to use herbal medicine research to expand the influence of their culture's indigenous herbal practices in the global healthcare market.

This illustrates that the perceived need for research may understandably differ from country to country and indeed between regions or continents. But to ascertain the impact of research, there should be areas of basic agreement on the primary source of social value for the research. In fact, the importance of partnering across expertise and varied capacities is crucial before a study on TM is decided. The requirements, though simple, involve thorough deliberations to discuss in detail the potential differences about the perceived need for the research and compatibility for research partnership based on the outcomes of social value assessments.

Therefore, studying available national, regional, and global policies and strategic plans and frameworks such as the Sustainable Development Goals (SDGs) for evidence-informed policy formulation and implementation makes it necessary and imperative to consider the roles and functions of policy actors and key local stakeholders, on TM policy issues in the region.

6.6 Conclusion

The first definition of traditional medicine was by the Expert Committee of the WHO Regional Office for Africa (WHO AFRO) in 1976 as the sum total of the knowledge, skills, and practices based on the theories, beliefs, and experiences indigenous to different cultures, whether explicable or not, used in the maintenance of health as well as in the prevention, diagnosis, improvement, or treatment of physical and mental illnesses. The general acceptance of the Sustainable Development Goals (SDGs), particularly by African countries, is evident in the alignment of the AU 2063 framework. Of specific importance is SDG 3 on good health and well-being. One of its thirteen targets is to achieve universal health coverage, defined as ensuring that all people have access to the needed health services of sufficient quality to be effective while ensuring that the use of these services does not expose the user to financial hardship.

Therefore, the current use of African traditional medicines (ATMs) and associated expenditures in seeking care from traditional health practitioners (THPs) compels governments and policymakers as well as key stakeholders and actors such as the private sector and citizens to re-evaluate the place of TM in contributing towards achieving UHC in African countries and prioritise the development of guidelines for THPs. Since over 60% of people in sub-Saharan Africa (SSA) live in rural areas where conventional or modern health care is not accessible and, even where accessible, is of low quality, not affordable, and sourced from weak health systems, exploring the role of ATM to achieve the goals of UHC becomes important. Furthermore, given the economic reality and cultural beliefs and with reference to the WHO

Beijing Declaration on Traditional Medicine of 2008, empowering traditional health practitioners (THP) will also enable more people to access affordable and regulated research-based TM.

Interestingly, TM continues to evolve and remains resilient in spite of the systematic neglect and the much more standardised Western or modern medicine. The neglect is attributed in part to the lack of or minimal documentation, but despite this, and as discussed earlier in this chapter, TM is still practiced, especially in rural communities. The existence of policy documents, strategic plans such as the WHO TM Strategic Plan, and other national and regional TM policies, plans, and frameworks, as well as the progress made toward integrating TM and modern medicine, presents the foundation for policy making that is inclusive of addressing policy implementation. While considerable debates seek to place a distinction between both TM and modern medicine, the focus should be on integrating both, drawing on the strengths and benefits as well as their complementarity towards improving the population's health and the well-being of the region.

A synopsis in the article entitled 'Traditional Medicine versus Modern Medicine' published on the Indonesia International Institute for Life Sciences (i3L) discusses the complementarity of traditional medicine and modern medicine and shows that while TM includes the biological, social, spiritual, and psychological dimensions to health issues, modern medicine often ignores or compartmentalises these aspects to fields such as mental health, physiotherapy, and the like. The high point of this debate agrees that despite the effectiveness of modern medicine, there are still crucial aspects of TM that can lead to successful treatment or the wellness of the patient. The goal is not to determine which one is better than the other because each has advantages and disadvantages of its own, but the main consideration is about the complementarity that will lead to the achievement of the same goal of improved well-being of mankind.

To this end and in view of the challenges that have emerged with the progressive evolution of TM over the years, the question of research is to ascertain, as much as possible, the efficacy, efficiency, and overarching benefits of TM (Kasilo et al., 2009) This should be done for TM policy formulation and practice. In view of the discovery of new and reoccurring epidemics, and even more recently, the COVID-19 pandemic, the human population must begin to explore new, effective, and efficient ways that also leverage on the current medical infrastructure designed to deal with some illnesses and infectious diseases while removing any basis for dispute in TM practice and utilisation in the West Africa sub-region.

References

1. EB124.R9 (2009) Traditional Medicine. Executive Board and World Health Assembly Resolutions on Traditional Medicine. https://apps.who.int/gb/ebwha/pdf_files/EB124/B124_R9-en.pdf (Accessed 13 February 2022).
2. Fokunang CN, Ndikum Yabi OY, et al. (2011). Traditional Medicine: Past, Present and Future Research and Development Prospects and Integration in the National Health System of Cameroon. Africa Journal of Traditional, Contemporary and Alternative Medicine (AJTCAM). Published online 2011 Apr 2. doi: 10.4314/ajtcam. v8i3.65276. (Accessed 27 March 2022)
3. Gyasi RM, Afriyie PA, Boateng S, et al. (2017). Integration for coexistence? Implementation of intercultural health care policy in Ghana from the perspective of service users and providers,, Journal of Integrative Medicine 15(1):44–55, DOI:10.1016/S2095-4964(17)60312-1 (Accessed 12 April, 2022)
4. Indonesia International Institute for Life Sciences (i3L), Traditional Medicine versus Modern Medicine. https://i3l.ac.id/traditional-medicine-vs-modern-medicine/. (Accessed 6 April 2022)
5. Kasilo OMJ et al. (2019). Towards universal health coverage: advancing the development and use of traditional medicines in Africa. BMJ Global Health 2019;4:e001517. doi:10.1136/bmjgh-2019-001517
6. Nadine I, Boon H (2018). Statutory Regulation of Traditional Medicine Practitioners and Practices: The Need for Distinct Policy Making Guidelines. Journal of Alternative and Complementary Medicine, 24(4): 307–313. DOI: 10.1089/acm.2017.0346.

https://www.ncbi.nlm.nih.gov/pmc/articles/PMC5909079/pdf/acm.2017.0346.pdf

7. Novartis Invests $100 Million in Shanghai. https://www.science.org/doi/10.1126/science.314.5802.1064b - Science. 17 Nov 2006. Vol 314, Issue 5802, pp. 1064-1065. DOI: 10.1126/science.314.5802.1064b. (Accessed 12 April, 2022)

8. Oyebode O, Kandala N, Chilton PJ et al. (2016). Use of traditional medicine in middle-income countries: a WHO-SAGE study. https://www.ncbi.nlm.nih.gov/pmc/articles/PMC5013777/. (Accessed 13 February 2022)

9. Pan S, Litscher B, Gao, Zhou S et al. (2014). Historical Perspective of Traditional Indigenous Medical Practices: The Current Renaissance and Conservation of Herbal Resources. (Accessed 11 July 2022)

10. Sato A (2012b). Revealing the popularity of traditional medicine in light of multiple recourses and outcome measurements from a user's perspective in Ghana. Health Policy and Planning 27: 625–37.

11. SciDevNet (2010). Integrating modern and traditional medicine: Facts and figures. https://www.scidev.net/global/features/integrating-modern-and-traditional-medicine-facts-and-figures/ (Accessed 11 April 2022)

12. Smith A, Jogalekar S et al. (2014). Regulation of natural health products in Canada. Journal of Ethnopharmacology, 158: 507–510.

13. Stafford GI, Pedersen ME, van Staden J, et al., (2008). Review on plants with CNS-effects used in traditional South African medicine against mental diseases. Journal of Ethnopharmacology, 28(3):513–537. (Accessed 12 April, 2022)

14. Tilburt JC, Kaptchuk TJ (2008). Herbal medicine research and global health: an ethical analysis. Bulletin of the World Health Organization 86: 594–9.

15. Abdullahi AA (2011). Trends and Challenges of Traditional Medicine in Africa. African Journal of Traditional, Complementary and Alternative Medicine, 8(5 Suppl): 115–123. Published online 2011 Jul 3. doi: 10.4314/ajtcam.v8i5S.5

16. van Weel C, Kidd MR (2018). Why strengthening primary health care is essential to achieving universal health coverage.

Canadian Medical Association Journal, 190(15): E463–E466. doi: 10.1503/cmaj.170784. https://www.ncbi.nlm.nih.gov/pmc/articles/ PMC5903888/. (Accessed 27 March 2022)

17. Wachtel-Galor S, Benzie IFF (2011). Herbal Medicine: An Introduction to It's History, Usage, Regulation, Current Trends, and Research Needs. Taylor & Francis.

18. World Health Organization (2021). Primary Health Care. https:// www.who.int/news-room/fact-sheets/detail/primary-health-care. (Accessed 11 April 2022).

19. World Health Organization (2004). WHO welcomes South Africa's commitment to Traditional Medicine. https://www. afro.who.int/news/who-welcomes-south-africas-commitment-traditional-medicine. (Accessed 27 March 2022)

20. WHO Traditional Medicine Strategy 2014–2023. https://apps. who.int/iris/rest/bitstreams/434690/retrieve. (Accessed 3 March 2022).

21. World Health Organization (2000). General Guidelines for Methodologies on Research and Evaluation of Traditional Medicine. https://apps.who.int/iris/bitstream/handle/10665/66783/ W?sequence=1. (Accessed 13 February 2022)

22. World Health Organization (2017). International Regulatory Cooperation for Herbal Medicines (IRCH). https://www.who. int/initiatives/international-regulatory-cooperation-for-herbal-medicines. (Accessed 4 May 2022)

23. World Health Organization (1986). The Ottawa Charter for Health Promotion. https://www.who.int/teams/health-promotion/ enhanced-wellbeing/first-global-conference . (Accessed 11 April, 2022)

24. WHO Beijing Declaration on Traditional Medicine (2008). https://www.who.int/director-general/speeches/detail/address-at-the-who-congress-on-traditional-medicine. (Accessed 11 April, 2022).

25. Yan Z, Feng Z, Jiao Z et al. (2022). Safety of Using Traditional Chinese Medicine Injections in Primary Medical Institutions: Based on the Spontaneous Reporting System 2016–2020 in Henan Province, China. Front. Pharmacol., 12 April 2022 | https://doi. org/10.3389/fphar.2022.761097. (Accessed 16 April, 2022).

7

Regulation of Herbal Medicines, Traditional Health Practitioners, and Traditional Medicine Practices in Africa

Julius Ossy Muganga Kasilo[1], Peter Bai James[2], Kofi Busia[3]

[1]*Traditional Medicine, Medicines Supply, Health Infrastructure, Equipment Maintenance including Health Technologies, Medicines and Traditional Medicine Unit, WHO/AFRO*
[2]*National Centre for Naturopathic Medicine, Faculty of Health, Southern Cross University, Australia.*
[3]*Principal Postgraduate Supervisor, Faculty of Medicine, Lincoln University College, Malaysia*

7.1 Introduction

In the pre-colonial era, traditional medicine was the only health system in African countries. Under colonial rule, traditional medicine practice was equated with witchcraft and regarded as contrary to the cause and ideals of the pre-eminent colonial religion and conventional medicine.

Though actively suppressed, it was still practiced, but in a less explicit manner. After independence, between the 1950s and 1960s, the trend to strengthen national and cultural identities began re-emerging, and currently, in many low- and middle-income countries, traditional medicine continues to be a significant source of primary health care for a large proportion of the population because of its cultural acceptability, affordability, and accessibility. In the last few years, there has also been an upsurge of interest in the use of traditional medicine in high-income countries, where it is usually referred to as complementary and alternative medicine.

The Alma Ata Declaration (1978) made by the International Conference on Primary Health Care was an important positive shift for traditional health care as it recognised for the first time the role of traditional medicine and its practitioners in primary health care (WHO, 1978). This has contributed to the growing international popularity of TM, creating benefits and opportunities for TM users and indigenous knowledge (Mbatha et al., 2012; WHO, 2013; le Roux-Kemp, 2010; and Abrams et al., 2020). Consequently, conventional health practitioners (CHPs) are increasingly reaching out for assistance from traditional health practitioners (THPs), especially in sub-Saharan Africa where communicable and non-communicable diseases have increased mortality and morbidity rates (Mbwambo et al., 2007). To date, millions of people across the globe continue to utilise THPs within primary health care (WHO, 2013; and Azhar et al., 2018), tapping the resourcefulness of THPs, who had been previously underutilised by the health systems (Homsy et al., 2004; Madiba, 2010; and James et al., 2018). This underutilisation is put into perspective by an argument suggesting that in sub-Saharan Africa, an estimated 40,000 patients are treated by one CHP, and one THP treats 500 patients, indicating the abundance of THPs over CHPs and how overwhelmed CHPs are (Mngqundaniso and Peltzer, 2008; and Nzimande et al., 2021). Collaboration between THPs and CHPs has been strengthened in some countries (Busia et al., 2010; and Madiba, 2010), whereas in others, the capacities of THPs have been strengthened, thus contributing to building stronger healthcare systems.

African governments have made commitments to integrate traditional medicine into the existing health systems as spelt out in the plans of action to implement the Lusaka and Windhoek declarations of the summit of heads of state and governments on the first (2001–2010) and second (2011–2020) decade of African traditional medicine (WHO, 2021; African Union, 2005). However, this commitment must be translated into action through mobilisation and allocation of adequate financial resource for effective implementation of national strategic plans. Following the colonial period, the organisational relationship between traditional medicine and conventional health systems in most African countries was a 'tolerant' one as THPs were free to practice so long as they did not claim to be registered medical practitioners. However, most countries are moving towards integrating traditional medicine into their public health systems.

Despite their general acceptance and use over the years, the safety of herbal medicines cannot always be guaranteed. Aside from the fact that all medicines have the potential of causing side effects, information on African traditional medicine use is scanty because of limited documentation. This reality justifies the critical importance of regulating herbal medicines. It has been observed that many of the problems associated with the use of herbal medicines arise mainly from the classification of many of these products as food, dietary supplements, or herbal medicines together with their inappropriate preparation or production. In several countries, quality tests and production standards tend to be less rigorous or controlled, and evidence of the quality, efficacy, and safety of these products is not always required before marketing.

In addition, in several cases, THPs are not certified or licensed. Their lack of expertise in certain areas and, in some cases, uncontrolled actions of charlatans have impacted negatively on the image and credibility of traditional medicine. The slow pace at which regulations are adopted and laws are passed by relevant national authorities has led to weak regulatory regimes and law enforcement, yet the regulation of THPs and their practices and herbal medicines is key to the promotion, development, and integration of traditional medicine into national health systems._Owing to the complexity of traditional medicines, it is essential that the products and practice be subjected to rigorous scientific evaluations and regulatory assessments, just as is done with conventional medical practices and medical products to guarantee the safety, quality, and efficacy of the practices and the respective products (WHO, 2000).

This chapter discusses the policy context for the regulation of herbal medicines, including those related to quality, safety, and efficacy assessment; product registration; marketing, distribution, and post-marketing surveillance; traditional medicine practitioners and their practices, skills, and qualifications required; and the challenges posed in attempts to regulate the sector.

7.2 The policy context for traditional medicine regulation

In considering the policy context for the regulation of traditional medicine in the WHO African Region, the three areas that are addressed are the regulation of traditional medicine products, hereafter called herbal medicines; the regulation of THPs; and the regulation of the practice of traditional medicine. National policies are the basis for defining the role of traditional medicine in national healthcare programmes, ensuring that the necessary regulatory and legal mechanisms are geared for promoting and maintaining safety and efficacy. In addition, such policies provide a legal basis for equitable access to healthcare resources and information about those resources.

Herbal medicines should meet the quality, efficacy, and safety standards set by the appropriate national authority before they are administered to patients or sold to the public. The World Health Assembly Resolution WHA62.13 on Traditional Medicine of 2009 (WHO, 2009) recognised that member states have different domestic legislations, approaches, regulatory responsibilities, and delivery models related to primary health care and urged member states, to formulate national policies, regulations, and standards, as part of comprehensive national health systems. The aim is to promote the appropriate, safe, and effective use of traditional medicine and to 'consider, where appropriate, establishing systems for the qualification, accreditation or licensing of traditional medicine practitioners and to assist them to upgrade their knowledge and skill in collaboration with relevant health providers'.

Similarly, at the regional level, resolutions AFR/RC50/R3 (WHO, 2000) and AFR/RC63/R6 (WHO, 2013) adopted by the WHO Regional Committee for Africa in 2000 and 2013, respectively, urged member states to, among other things, 'strengthen the regulation of traditional medicine practitioners, practices, and products, including advertising, and protect the public against quack practitioners and illicit products'. The resolutions further urged member states 'to strengthen the capacity of national medicines regulatory authorities to issue marketing authorization for traditional medicine products that meet national criteria and WHO norms and standards of quality, safety, and efficacy and to undertake joint reviews of traditional medicine products registration files. The Resolutions requested WHO to provide technical support to strengthen national medicines regulatory authorities with a view to enhancing cooperation in and harmonization of the regulation of traditional medicine practitioners, practices and products'.

WHO has provided three types of support to countries in these areas. The first is direct technical support for the development of national tools or technical documents that will empower countries to establish professional regulatory bodies and enforce regulation for TM practitioners and practices as indicated above. The second is technical (and financial) support for the establishment of the THPs' council and its composition, functions, and registration, as well as the licensing of THPs. The third type of support is the provision of tools and guidelines developed by WHO for adaptation to specific contexts to assist countries accelerate the process of regulating THPs. These tools and guidelines include the *Model Legal Framework for the Practice of Traditional Medicine: A Traditional Health Practitioners* Bill (WHO, 2004), which provides for the establishment of the THPs' council or board and spells out its functions, among other important matters, and the *Framework for* Regulation *of Traditional Medicine Practitioners,*

Practices and Products (WHO, 2016d), which provides information on elements to regulate practitioners (practices and products) as well as requirements for licensing of practitioners and their premises.

7.2.1 Traditional health practitioners: A traditional health practitioner (THP), traditional healer, or traditional medicine practitioner (TMP) is a person recognised by the community in which he/she lives as competent to provide health care by using vegetable, animal, and mineral substances and certain other methods based on their social, cultural, and religious background as well as on the knowledge, attitudes, and beliefs that are prevalent in the community regarding physical, mental, and social well-being and the causation of disease and disability (WHO, 1976).·

Traditional health practitioners were practicing long before conventional medicine was introduced to middle- and low-income countries (Hoff, 1997). They provide up to 90% of primary health care (PHC) for people living in rural areas in many countries in the African region. THPs are involved in various specialisations and attempts have been made to categorise them based on their method of healing or the ailments they treat. One categorization, based on the type of healing, divides THPs into herbalists, ritualists and spiritualists, traditional midwives, traditional birth attendants, psychiatrists, and traditional bone setters, among others. It then subdivides these categories into generalists and specialists according to their degree of polarisation in treating one or more illnesses. For instance, a traditional bone setter is a lay practitioner of joint manipulation. He or she is the 'unqualified practitioner' who takes up the practice of healing without having had any formal training in accepted medical procedures; however, he or she will have received an apprenticeship.

An overview of the situation of African traditional medicine has been outlined in *African Indigenous Medical Knowledge and Human Health* by Kasilo et al. (2018), whereas Kasilo and Wambebe (2021) highlighted the importance of traditional and complementary medicine in global health care. In addition, Kasilo et al. (2019) highlighted how African traditional medicine and THPs could make significant contributions to the attainment of universal health coverage (UHC). WHO provided technical tools to assist African countries to develop African TM as a significant component of health care. Many African countries adopted the WHO tools after appropriate modifications to advance research and development (R and D) of African TM. An analysis of the extent of this development was done through a survey of forty-seven countries in the WHO African region. Results showed impressive advances in R and D of African TM, the level of collaboration between THP and conventional health practitioners (CHPs), quality assurance and regulation, registration, and THP integration into the national health systems.

However, the level of education of THPs varies. Some had primary school education, while others had undertaken secondary or high school, or even university education with degrees. Interestingly, others were very experienced medical doctors, who had decided to practice traditional medicine after many years of conventional medical practice. The degree of training and qualification of THPs also varied. Some had undergone vigorous and lengthy periods of training, while others had been initiated or called. Furthermore, other THPs had received their traditional medical knowledge from their forebears. Meanwhile, some THPs had little or no training and low ethical standards, lending them to derogatory descriptions.

7.2.2 Traditional medicine practice: Practices of traditional medicine vary greatly from country to country and from region to region as they are influenced by factors such as culture, history, personal attitudes, and philosophy. In many cases, their theory and application are quite different from those of conventional medicine. In general, knowledge of TM is generally passed on from generation to generation, and not systematically recorded and analysed, there exists some evidence of safety and efficacy of many herbs in particular. However, scientific research is still needed to validate ethnomedical uses, and provide additional evidence of its safety and efficacy. In undertaking this exercise, knowledge and experience obtained through the long history of established practices should be respected (WHO, 2000).

There are several domains of African traditional medicine practices, including general traditional health, mental health, traditional midwifery, curative, preventative, and surgical practices (Mhame et al., 2010). General health practices are services provided to clients by non-specialised healthcare providers, including THPs. The general THP manages conditions such as malaria, stomach infections, respiratory problems, rheumatism, arthritis, sexual dysfunction, anaemia, and parasitic infections. In mental health practices, THPs make use of divination to unravel the mental and psychological problems of their patients. Divination plays a significant role in the treatment of neurosis and helps retrace a patient's life from its metaphysical past to how it interplays with the present and future (Agarwal and Agarwal, 2010). In traditional midwifery practices, healthcare providers give prenatal care to pregnant women (low-risk pregnancy), assist in childbirth, and provide postpartum care to the mother and child.

In curative practices, THPs use plants as their primary means of providing treatment for various ailments and diseases. Preventative practices include several forms of disease prevention. These can include avoiding places where epidemics occur, inoculating people with pus from a sick person during special rituals, sweeping or covering floors with particular plants, isolating people with contagious diseases, prohibiting or controlling movement, and taking children away from affected areas. Surgical practices include bone setting, uvulectomy, circumcisions, bleeding and cupping, cautery, scarification, and tooth extraction. These practices are recognised by communities and some governments, but other practices such as divination and circumcision are not always accepted.

7.3 Regulation of herbal medicines

Herbal medicines are usually preparations of one or more herbs. If more than one herb is used, the term 'herbal mixture' is used. Finished herbal products and herbal mixtures may contain excipients, in addition to the active ingredients. However, finished products or herbal mixtures to which chemically defined active substances have been added, including synthetic compounds and/or isolated constituents from plant materials, are not considered to be herbal medicines.

Regulation of herbal medicines is a key requirement, consistent with the general regulation of medical products. Herbal medicines are regulated by national medicines regulatory authorities (NMRAs), which issue a marketing authorisation (product licence, registration certificate) that authorises the marketing or free distribution of a herbal medicine in the respective country after evaluation for safety, efficacy, and quality.

To promote the registration, distribution, and use of effective, quality and safe traditional medicines in countries, the WHO Regional Office for Africa in 2004 issued guidelines, which were reprinted in 2010 (WHO, 2010). The guidelines contain a classification of traditional medicines as indicated above, and minimum regulatory requirements for their registration vis-à-vis determination of quality, safety, and efficacy by national drug regulatory authorities. Similar guidelines, protocols, or regulatory frameworks (WHO, 2016a) have also been developed for assessing the safety, efficacy, and quality of traditional medicines and accelerating the protection of traditional medical knowledge and intellectual property rights (WHO, 2016b, 2016c).

In terms of quality, NMRAs establish inter alia the detailed composition and formulation of the herbal medicine, and the quality requirements for the product and its ingredients. It also includes details of the packaging, labelling, storage conditions, shelf life, and approved conditions of use. Despite this, most countries in the region have not yet established safety monitoring mechanisms for imported and locally produced herbal medicines, as demonstrated by surveys conducted by WHO between 2000 and 2020, which showed that only four and twenty-three countries, respectively, had reviewed their essential medicines regulations to include the registration of herbal medicines. This seemed to reflect the inadequacy of facilities for researchers in the region to assess the quality, safety, and efficacy of herbal medicines whose composition is usually complex.

7.3.1 Assessment of quality, safety, and efficacy: The establishment of quality is an indispensable process in the production of any therapeutic agent. Proper identification of a medicinal plant material is fundamental to the quality control process, as it must be established unequivocally that the source of the plant material is authentic. Ethnobotany and pharmacognosy are effective tools for achieving this. Following this, contamination (fungal and bacterial) must be checked during processing of the plant material. Chemical, microbiological, pharmacological, and toxicological evaluations conducted according to the principles of good laboratory practices (GLPs), will help to certify the bioactive properties of the material undergoing processing (WHO, 2007). The provisions in the guidelines range from raw plant materials, through processed, packaged remedies, to imported herbal products. The guidelines can be used to determine the kind of product to be made even before the product is manufactured. The chemical, pharmacological, and toxicological evaluations also are often the predictors of safety of the products manufactured.

Clinical safety and efficacy will need to be established through expensive and usually lengthy clinical trials during the development of a therapeutic agent. If the safety and efficacy profiles obtained from randomised controlled clinical trials are of acceptable standard, then the dossier, along with other documents, can be submitted to the national medicine regulatory authority for registration. After that, so long as the standard operating procedures are adhered to, then the unit dosage forms produced will be considered safe. Notwithstanding this, quality assurance procedures must be instituted so that the products coming from the factory are of consistently good quality, safety, and efficacy.

It is usually difficult for THPs to prove the efficacy of their products, and research institutions do this in some countries such as Ghana and Mali, where such arrangements are made. The products are therefore regarded as complementary medicines, and all that may be required is proof of safety and quality.

7.3.2 Safety monitoring: this is a fundamental principle in the provision of traditional medicines for health care and a critical component of quality assurance. However, the safety monitoring of herbal medicines in Africa is not carried out routinely. Adverse reactions of herbal medicines are difficult to distinguish from those attributable to the poor quality of medicinal products or due to inappropriate use of the wrong species of medicinal plants or unsafe, irrational, and improper use of herbal medicines. In addition, there is a general lack of knowledge of herbal medicines by conventional health practitioners and consumers. Therefore, awareness will need to be created through education and communication so that all providers and consumers are well-informed about the potential adverse effects of herbal medicines. WHO has developed guidelines on safety monitoring to assist member states in their bid to produce safe herbal medicines (WHO, 2004).

The basic procedure of registration of medicines, whether traditional or conventional, is the same as shown in Figure 1, and the *Guidelines on Registration of Traditional Medicines in the WHO African Region* will serve as a reference guide for the registration of herbal medicines.

National regulatory authorities are a precious resource for investigators, healthcare professionals, and industries to facilitate the application of good practice requirements to herbal medicines and build the capacity of THPs, the THP Council, and their associations in the appropriate methodologies of cultivation, research, production, storage, distribution, and dispensing of plant-based products. Collaboration with associations and federations of THPs should be institutionalised and documented through memoranda of understanding or other appropriate mechanisms, which would spell out the roles of these institutions in

- disseminating good practices and monitoring their application,
- facilitating access to quality control laboratories for raw materials and product testing,
- collecting and evaluating adverse drug reaction reports,
- facilitating links between THPs and pharmacists to guide rational use of medicinal plants, and
- creating the infrastructure necessary to ensure the quality of products, including those produced at a small scale.

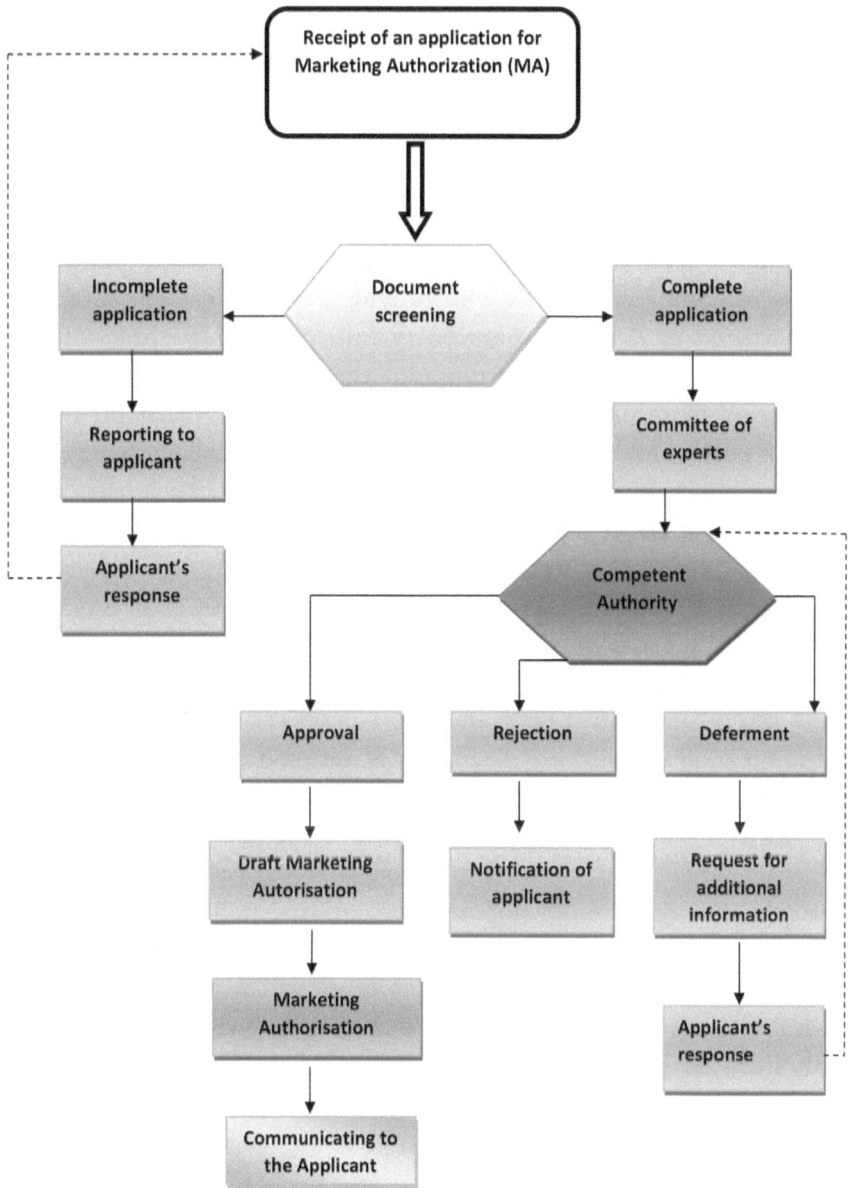

Figure 1: Flow chart of basic procedure of registration of a herbal medicine leading to granting of marketing authorization by a National Regulatory Authority in the African Region (WHO/AFRO Regional framework for regulation of traditional medicine practitioners, practices, and products (document AFR/EDM/HTI/2016.5, WHO, 2016d)

7.3.3 Regulation of traditional health practitioners and their practices

Regulation of THPs is a principle, rule, or directive designed to control or govern the conduct and practice of these practitioners, who are obliged to adhere to a specific piece of legislation. The objectives of regulating THPs are to, among others, ensure that the public has access to competent THPs who provide safe and ethical care, and are certified or licenced upon entry into professional practice and maintain high standards throughout their active careers. They are also meant to ensure that practitioners adhere to and maintain a good code of ethics and practice, particularly in relation to their work, their patients, their colleagues, and the general public; make provision for the control of the registration, training, and practices of THPs; and serve and protect the interests of members of the public who use the services of THPs and thereby curtail the practice of charlatans and unqualified practitioners.

It is important that THPs and their practices are regulated by a national professional body such as a traditional health practitioner's council (THPC) established within the ministry responsible for health, just as it is done in other professions. For the full potential of TM to be realised, there is a need to officially recognise its role in health systems through the development of national policies. THPs need to be empowered with the requisite regulatory and legal framework. The framework should include a code of ethics and practice, and minimum requirements for the practice of traditional medicine so that only licenced THPs would be allowed to practice as provided for in the WHO *Tools for Institutionalizing Traditional Medicine in Health Systems*.

In addition, such a legal and regulatory framework could assist THPs to organise themselves into functional associations or federations and councils for effective policy implementation. Membership of such associations or federations should be based on accreditation, registration, or licencing of qualified practitioners to help eliminate quackery and charlatanism. Such policy orientations could radically enhance traditional medicine development in the African region. To date, all countries in the region have established associations of THPs, and more than half have umbrella national associations of practitioners, and some associations such as in Mali have formed federations.

In regulating THPs, countries need to consider the licencing of THPs and facilities for their practice, and ensure their adherence to the code of ethics. The regulatory inspections of premises where all medicines, including herbal medicines, are handled are carried out by national regulatory authorities (NRAs). These authorities and THP councils/boards/commissions need to work together to ensure that this is implemented. The details on the elements to regulate in each of these aspects are contained in the regional framework for the regulation of traditional medicine practitioners, practices, and products.

7.4 Qualification and skills of THPs and their registration criteria: THPs should meet the qualification criteria for the practice of TM, as determined by the professional regulatory body of THPs. The qualification criteria should include years of experience, level of education, and the process for becoming a THP, such by calling, apprenticeship, or through inheritance from a family member, as well as the opinions of both the community and that of the local government where the practitioner resides and the findings of surprise visits to the practitioner's practice facility.

In addition, the legal framework must be adopted by the national cabinet, passed into law and reinforced by the THP's council/board/commission. To accelerate the pace of law enforcement and implementation of some of its functions, the council/board/commission should establish committees as required, consisting of members of the council and non-members, to exercise any of its functions. For example, the committees may include a registration committee which would collaborate with associations of THPs to support the THP council on all issues related to the identification of qualified practitioners using specific criteria, as well as deal with issues related to the licencing of qualified THPs. The research and training committee would, for example, promote and support education, training, and research in traditional medicine, promote the practice of THPs, and develop educational and training methods.

The skills required of THPs for the practice of traditional medicine vary from country to country. However, in some countries, the minimum requirements are as listed in Box 1. Similarly, standards for the practice of traditional medicine in terms of qualifications also vary from country to country. Some of the minimum skills required of THPs for the practice of traditional medicine, which countries may wish to consider adopting, include those listed in Box 2. However, the minimum standards indicated in Boxes 1 and 2 shall only serve to guide each THP's council in establishing national standards.

The registration system for THPs will help to reduce the widespread sale of fake herbal medicines in villages and rural communities with unsubstantiated claims, because only certified THPs will be permitted to market their products.

7.5 Code of ethics and practice for traditional health practitioners: To support countries and accelerate the process of regulating traditional medicine practice, the WHO has developed *Model Codes of Ethics and Practice for Traditional Health Practitioners*, which contains four important areas to be observed by THPs. These are the *Model Code of Ethics, the Model Code of Practice, a Model for Disciplinary Procedures*, and *a Model for Minimum Standards for THPs*. A code of ethics is a set of rules or moral guidelines that define and govern principles and actions in an organisational environment. Generally, a code of ethics promotes an environment of respect based on integrity. When people know the code of ethics and follow it, this creates an atmosphere of trust, mutual respect, and integrity, as well as confidence in the actions of each person involved in the organisation or group. Ethical behaviour is beneficial for everyone because it protects the interests of the association or federation of THPs and the interests of everyone who meets the association or federation.

The *Model Code of Ethics and Practice for Traditional Health Practitioners* seeks to ensure a high standard of conduct and practice among THPs; foster good relationships among them practitioners, and their patients and other health professionals; and increase THPs' awareness of the existing rules and regulations governing the practice. With the existence of a written code of ethics, THPs, will be expected to behave professionally among themselves, towards conventional health practitioners, and towards their patients.

The *Model Code of Practice* also includes the requirements of the code of practice THPs must adhere to. These may include the premises and location of the practice, application for licence, areas of competence, patient examination and treatment, fees to be charged, patient records and notifiable diseases, and handling/storage, labelling, and administration of traditional medicines. The *Model Code of Practice* further spells out the disciplinary measures to be applied in the event of professional misconduct. *A Model for Disciplinary Procedures* stipulates the disciplinary procedures to be followed by the appropriate national authority in the event of professional, dishonourable, or ethical misconduct.

To facilitate the regulation of traditional health practitioners and their practices, each country should develop and:

- implement a national traditional medicine policy, which provides direction on the development and practice of TM in that country;
- enforce a legal framework for the practice of TM through the regulation of THPs by a functional THP's council/board/commission in accordance with the national THPs Act; and
- enforce a code of ethics and practice by a functional THP's council/ board/commission to ensure the safety, efficacy, and quality of the services provided by a qualified and licenced THP.

Box 1: Minimum Qualification Requirements of Traditional Health Practitioners	Box 2: Minimum Skills Required of Traditional Health Practitioners
The minimum level of education for all traditional health practitioners should be the primary school leaving certificate.	Each traditional health practitioner's council, board, or commission should determine the skills required for practicing traditional medicine.
A traditional health practitioner must be a member of a recognised traditional health practitioners association or federation in the community where the traditional health practitioner practices.	After registration and licencing by a professional regulatory body such as a council, board, or commission, a general herb seller must be able to recognise at least fifty (50) herbs, while those selling herbs for specific ailments must be able to identify a minimum of ten (10) herbs that can be used to treat various ailment(s).
Every traditional health practitioner must have successfully undergone the certification procedure stipulated for registration by the competent national authority/the THP's board, council, or commission.	A traditional health practitioner must be able to recognise at least thirty (30) different herbs.
	A traditional health practitioner who administers herbs should be able to submit at least two medicaments prepared and used in practice so that they can be tested for efficacy (patent rights agreements should be signed as appropriate) and safety by a designated research institute. The state of hygiene of the THP's facility should be approved by the national regulatory authority.
	Practitioners who are traditional surgeons should limit themselves to the practice of non-invasive surgery, i.e. surgery not beyond the epidermis of the skin.
	All traditional health practitioners, regardless of their mode of practice or their skills, must conduct their practice within the limits of the law of their countries.

7.5.1 Case study 1: Provides an example of the legislative landscape for traditional health practitioners in Southern African development community countries.

Case study 1: An example of the legislative landscape for traditional health practitioners in Southern African Development Community countries

A scoping review on the legislative landscape for traditional health practitioners in Southern African Development Community (SADC) countries was carried out, with the aim of mapping and reviewing THP-related legislation among SADC countries. To limit the scope of the review, the emphasis was on defining traditional health practitioners (THPs) in terms of legal documents.

It was found that four Southern African countries (Namibia, South Africa, United Republic of Tanzania, and Zimbabwe) of fourteen have legislation relating to THPs. Three of these countries, Namibia, South Africa, and Zimbabwe, acknowledged the roles and importance of THPs in healthcare delivery by creating a THP's council to register and formalise practices, although they had not operationalised or registered and defined THPs. In contrast, the United Republic of Tanzania has established a definition couched in terms that acknowledge the context-specific and situational knowledge of THPs while also outlining methods and the importance of local recognition. Tanzanian legislation thus provides a definition of THP that specifically operationalises THPs, whereas legislation in Namibia, South Africa, and Zimbabwe allocates the power to a THP's council to decide or recognise who a THP is; this council can prescribe procedures to be followed for the registration of a THP.

The review highlighted the differences and similarities between the various TM policies and legislation in SADC countries. Legislation concerning THPs is available in four of the fourteen SADC countries. While Namibia, South Africa, United Republic of Tanzania, and Zimbabwe have legislation that provides guidance as to THP recognition, registration, and practices, THPs continue to be loosely defined in most of these countries. Not having an exact definition for THPs may hamper the promotion and inclusion of THPs in national health systems, but it may also be unavoidable, given the tensions between lived practices and rigid legalistic frameworks.

7.5.2 Case study 2 provides an example of the regulation of traditional medicine practice in Ghana, extracted from the Traditional Medicine Practice Act 575 (2000)

Case study 2: An example of the regulation of traditional medicine practice in Ghana

The Traditional Medicine Practice Act 575 (2000) in Ghana provides for the establishment and functions of the Traditional Medicine Practice Council (TMPC), registration of practitioners, qualification for registration, and licensing of practices. It establishes the council to regulate the practice of traditional medicine, register practitioners and licence practices, regulate the preparation and sale of herbal medicines, and provide for related matters. The council shall have perpetual succession and a common seal and may sue and be sued in its own name. The council may, for the performance of its functions, acquire and hold movable and immovable property and may enter into a contract or any other transaction.

The objective of the Traditional Medicine Practice Council is to promote, control, and regulate traditional medicine practice. The council shall

- set standards for the practice of traditional medicine;
- issue a certificate of registration to a qualified practitioner and licence premises for a practice;
- determine and enforce a code of ethics for traditional medicine practice in conjunction with an association of traditional medicine practitioners recognised by the Minister of Health;
- promote and support training in traditional medicine;

- approve, in consultation with the educational and research institutions determined by the board, the curriculum for training in traditional medicine in institutions;
- collaborate with the ministry to establish centres for provision of traditional medical care within the national healthcare delivery system;
- advise the minister on matters relating to and affecting the practice of traditional medicine; and
- collaborate with the appropriate agencies for large-scale cultivation of medicinal and aromatic plants and the preservation of biodiversity:

(i) advising the Food and Drugs Board in writing on the rules for the registration, advertisement, manufacture, packaging, preparation, labelling, sale, supply, exportation, and importation of herbal medicine; (ii) monitoring the fees payable by clients for services provided by practitioners; and (iii) performing any other functions that are ancillary to the objectives of the council.

The Traditional Medicine Practice Act 575 provides for the registration of practitioners and qualification for registration. A person shall not operate or own premises as a practitioner or produce herbal medicine for sale unless that person is registered in accordance with this act. A person seeking full or temporary registration shall apply to the registrar in the manner determined by the board.

Qualifications for registration:

- An applicant has adequate proficiency in the practice of traditional medicine.
- The application has been endorsed by any two of the following: the district chairman of the association, the traditional ruler of the community, and the district coordinating director – he/she shall direct the registrar to enter the applicant's name in the register of practitioners and issue the applicant with a certificate of registration on the payment of the prescribed fee by the applicant.

- A person issued with a certificate under subsection (I) shall be known as a practitioner for the purposes of this act.
- Registration under this act is in addition to the registration required under any other law in respect of the practice. With respect to licensing of practices, a person shall *not own or operate a practice unless that person holds a licence in respect of the practice* issued under this act.
- A person may apply to the council for a licence for a practice through the district office of the council within the area in which the practice is to be operated in the form determined by the board.

7.6 Challenges

There are many challenges in the development and implementation of policies for the regulation of herbal medicines, traditional medicine practitioners and practices. The principal ones are as follows:

- **Lack of financial resources to implement resolutions, declarations, and recommendations on traditional and complementary medicine made at the global, continental, and regional level:** The lack of political will and financial resources to implement current TM policies and frameworks is an overarching challenge.
- **Differences in regulatory status of herbal medicines:** There are great differences between countries in the definition and categorisation of herbal medicines. A single medicinal product may be defined as a food, a functional food, a dietary supplement, or a herbal medicine in different countries, depending on the regulations applying to foods and medicines in each country. This makes it difficult to define the concept of herbal medicines for the purposes of national medicine regulation and may confuse patients and consumers.
- **Difficulty in assessing safety and efficacy:** Herbal preparations may contain hundreds of active constituents, which will require a lot of time and resources to isolate.

Therefore, there is inadequate data on the scientific and clinical validation of many traditional medicines.

- **Difficulty undertaking quality control:** The safety and efficacy of herbal medicines are closely related to the quality of the source raw materials, which are determined by intrinsic factors (genetic) and extrinsic factors (environmental conditions, cultivation and harvesting, field collection and post-harvest/collection, transport, and storage). The practical approach for quality control is to identify a chemical and biological marker, which can be utilised for each production batch.

- **Difficulty morning safety:** Adverse events arising from consumption of herbal medicines may be due to factors, which may be exacerbated by the absence of robust regulatory systems. These factors include misidentification, adulteration, wrong labelling, contamination with toxic or hazardous substances, over dosage, misuse by either healthcare providers or consumers, and concomitant use with other medicines. Assessing the adverse events resulting from use of herbal medicines is therefore more complicated than usual.

- **Limited knowledge about herbal medicines within national medicine regulatory authorities**, resulting in use of appropriate evaluation methods.

- **Proliferation of practitioners with doubtful abilities and intentions:** According to Pretorius (1999) 'in the current economic climate and amid the high unemployment, there is a marked increase in the ranks of THPs, among whom there are, unfortunately, quite a number of charlatans'.

- **Mutual distrust between allopathic and THPs in Africa:** This has continuously hampered and thwarted the process of integration and cooperation between traditional and modern medicines) as well as the difficulties in regulating traditional medical practices. For instance, Ebomoyi (2009) reported that Nigerian medical students have reservations about integrating TM into mainstream healthcare provision in the country.

7.7 Conclusion

Traditional medicine still plays an important role in healthcare delivery in African countries. There are many actions that countries need to take to strengthen the capacity of member states for the regulation of herbal medicines and traditional medicine practitioners and practices to ensure quality, safety, and efficacy.

Member states need to develop national policies, strategies, and tools for regulating practitioners and their practice and products. The tools and guidelines developed by WHO and other partners can be adopted and adapted by countries to their unique circumstances. Countries need to strengthen the regulation of traditional medicine practitioners, and their practices products to prevent quackery. Countries also need to strengthen the capacity of professional associations and traditional medicine practitioners' regulatory bodies to identify qualified THPs for accreditation or licencing.

The capacity of national medicines regulatory authorities needs to be strengthened to enable them achieve their mandate of issuing marketing authorisations for herbal medicines that meet national criteria and WHO norms and standards of quality, safety, and efficacy. Government policymakers, regulators for THPs and their practice and products, and international development partners and donors need to understand the basic tenets of regulation and work together to build strong systems of accreditation, licencing/registration, and professional development. Furthermore, the general public should be educated to only consult qualified and licensed THPs, who will be recognised through the certificates posted in their facilities.

Despite the above-mentioned challenges, there are opportunities for improvement of the regulatory environment provided by the newly established African Medicines Agency (AMA), whose vision is to ensure that all Africans have access to quality-assured, safe, efficacious, and affordable medical products that meet internationally recognised standards for priority diseases or conditions. At the continental level, AMA's mission is to coordinate national and sub-regional medicines regulatory systems; conduct regulatory oversight of selected medical products, including traditional medicines; and promote cooperation, harmonisation, and mutual recognition of regulatory decisions. AMA holds promise to address gaps and inconsistencies in national regulatory legislation and ensure effective medicines regulation, including herbal medicines.

For its part, WHO and its partners, as well as regional economic communities, will continue to provide support to countries to address the gaps and challenges in the regulation of traditional medicine practitioners, practices and products. WHO will also provide support to strengthen national medicines regulatory authorities to enable them function effectively. In addition, WHO will create the platform for collaboration, experience sharing, dissemination of best practices, and harmonisation of the regulation of traditional medicine practice at regional and sub-regional levels.

References

1. Abrams A, Falkenberg T, Rautenbach C, et al. (2020). Legislative landscape for traditional health practitioners in Southern African development community countries: a scoping review. *BMJ Open.*;10:1. doi:10.1136/bmjopen-2019-029958.

2. African Union (2005). Plan of action on the African Union Decade of Traditional Medicine (2001–2010). Implementation of the decision of the Lusaka Summit of Heads of State and Government. (AHG/DEC.164 (XX^{XV}II). 2nd ordinary session of the conference of African Ministers of health (CAMH2), Gaborone, Botswana, 10–14 October 2005.

3. Agarwal A, Agarwal R (2010). The Practice and Tradition of Bone setting Education for Health, 23(1). Available from: http://www.educationforhealth.net/

4. Azhar H, Alostad AH, Steinke DT, Schafheutle EI (2018). International Comparison of Five Herbal Medicine Registration Systems to Inform Regulation Development: United Kingdom, Germany, United States of America, United Arab Emirates and

Kingdom of Bahrain. Pharmaceutical medicine, 32(1):39–49. doi:10.1007/s40290-018-0223-0.

5. Boly R, Compaore S, Ouedraogo S, Zeba M, et al. (2021) Collaboration between practitioners of traditional and conventional medicine: A report of an intervention carried out with traditional women healers in the province of Sanmatenga (Burkina Faso) to improve the obtaining of the license to practice traditional medicine. International NGO Journal, 16(1), 9–16. Available at: https://academicjournals.org/journal/INGOJ/article-full-text-pdf/E24D14566598 .

6. Busia K, Kasilo OMJ, Mhame PP (2010). Collaboration between traditional health practitioners and conventional health practitioners: some country experiences. Afr. health monit. (Online), 14: 40–46.

7. Ebomoyi EW (2009). Genomics in Traditional African Healing and Strategies to Integrate Traditional Healers into Western-Type Health Care Services- A Retrospective Study. Researcher, 1(6):69–79.

8. Pretorius E (1999). South African Health Review. 5th Edition. Durban: Health Systems Trust. "Traditional Healers" 249–256. WHO (1978). The Promotion and Development of Traditional Medicine. World Health Organization, 1–44.

9. Homsy J, King R, Balaba D, Kabatesi D (2004). Traditional health practitioners are key to scaling up comprehensive care for HIV/AIDS in sub-Saharan Africa. Aids, 18(12):1723–1725. doi:10.1097/01.aids.0000131380.30479.16.

10. Hoff W (1997). Traditional health practitioners as primary health care workers. Tropical Doctor, 27:suppl 52–55.

11. James PB, Wardle J, Steel A, et al. (2018). Traditional, complementary and alternative medicine use in sub-Saharan Africa: a systematic review. British Medical Journal. Global Health, 3:e000895.

12. Kasilo OMJ, Nikiema JB, Desta A. and Loua A (2018). The situation of traditional medicine in Africa. Where are we? In: Charles Wambebe (Eds). African Indigenous Medical Knowledge and Human Health, ISBN 9781032095806. Published by CRC Press, pp. 1–50.

13. Kasilo OMJ, Wambebe C. (2021) Traditional and Complementary Medicine in Global Health Care. In: Haring R., Kickbusch I., Ganten D., Moeti M. (eds) Handbook of Global Health. Springer, Cham. https://doi.org/10.1007/978-3-030-05325-3_63-1.

14. Kasilo OMJ, Wambebe C, Nikiema J-B, et al. (2019). Towards universal health coverage: advancing the development and use of traditional medicines in Africa. British Medical Journal. Global Health, 4:e001517. Doi:10.1136/ bmjgh-2019-001517.

15. le Roux-Kemp A (2010). A legal perspective on African traditional medicine in South Africa. *The Comparative and International Law Journal of Southern Africa.*;43(3):273–291.

16. Mbatha N, Street RA, Ngcobo M, Gqaleni N (2012). Sick Certificates Issued by South African Traditional Health Practitioners: Current Legislation, Challenges and the Way Forward. The South Medical Journal, 102, 3.

17. Madiba SE (2010). Are biomedicine health practitioners ready to collaborate with traditional health practitioners in HIV and AIDS care in Tutume sub district of Botswana. African Journal of Traditional, Complementary, and Alternative Medicines, 7(3):219–224.

18. Mbwambo ZH, A Mahunnah RL, A Kayombo EJ (2007). Traditional health practitioner and the scientist: bridging the gap in contemporary health research in Tanzania. Tanzania Health Research Bulletin.;92:115–200.

19. Mhame P, Busia K and Kasilo OMJ (2010) Clinical practices of African traditional medicine. Special Issue of African Health Monitor, 14, 31–39.

20. Mngqundaniso N, Peltzer K (2008). Traditional Healers and Nurses: A Qualitative Study on Their Role on Sexually Transmitted Infections Including HIV and AIDS in KwaZulu-Natal, South Africa. African Journal of Traditional, Complementary, and Alternative Medicines, 5(4):3

21. Nzimande SI, Moshabela M, Zuma T, Street R, Ranheim A, Falkenberg T (2021). Barriers and facilitators of traditional health practitioners' regulation requirements: a qualitative study. Journal of Global Health Reports, 5:e2021014. doi:10.29392/001c.21402.

22. Traditional Medicine Practice Act, 2000 ACT 575. Available at: https://wipolex-res.wipo.int/edocs/lexdocs/laws/en/gh/gh025en.html, accessed on 17 June 2022.

23. WHO (2021). African Traditional Medicine Day 2021. Message of WHO Regional Director for Africa, Dr Matshidiso Moeti. https://www.afro.who.int/regional-director/speeches-messages/african-traditional-medicine-day-2021. Accessed 8-01-2023.

24. WHO (2013). Traditional Medicine Strategy 2014–2023. Department of Essential Drugs and Medicines Policy, 78.

25. WHO (2000). General Guidelines for Methodologies on Research and Evaluation of Traditional Medicine (Document, WHO/EDM/TRM/2000.1). World Health Organization, Geneva.

26. WHO (2009). Resolution WHA62.13 on Traditional Medicine, World Health Organization, Geneva.

27. WHO (2000). Promoting the Role of Traditional Medicine in Health Systems: A Strategy for the African Region (Document AFR/RC50/9) adopted by its Resolution, AFR/RC50/R3 by the Fiftieth Session of the WHO Regional Committee for Africa. AFR/RC63/R6 on Enhancing the Role of Traditional Medicine in Health Systems: A Strategy for the African Region (Document AFR/RC63/6) by its Resolution, AFR/RC50/R3 by the Fiftieth Session of the WHO Regional Ie for Africa, adopted by its Resolution, AFR/RC63/R6 by the Sixty-Third Session of the WHO Regional Committee for Africa.

28. WHO (2004). Model Legal Framework for traditional medicine practice; A traditional Health Practitioners Bill. In: Tools for institutionalizing traditional medicine in health systems in the African Region. AFR/TRM/04.3, World Health Organization Regional Office for Africa, Brazzaville.

29. WHO (2016). Regional framework for regulation of traditional medicine practitioners, practices and products (document AFR/EDM/HTI/2016.5).

30. WHO (1976). Technical Report Series, No.1, African traditional medicine, Report of the Regional Expert Committee. WHO Regional Office for Africa, Brazzaville, pp 3–4.

31. WHO (2000). General Guidelines for Methodologies on Research and Evaluation of Traditional Medicine (Document, WHO/EDM/TRM/2000.1). World Health Organization, Geneva.

32. WHO (2004). Guidelines for regulation of traditional medicines in the WHO African Region. Reprinted in 2010 in India. WHO Regional Office for Africa, Brazzaville, Republic of Congo.

33. WHO (2016a). WHO Regional Regulatory Framework for Regulation of Traditional Medicine: Practitioners, Practices, and Products. WHO Regional Office for Africa, Brazzaville, Republic of the Congo (Document AFRO/EDM/TRM/2015.5). (NLM Classification: WB 55).

34. WHO (2016b). WHO Regional Policy Guidance on National Policy for the Protection of Indigenous Knowledge in African Traditional Medicine. WHO Regional Office for Africa, Brazzaville, Republic of the Congo (Document AFRO/EDM/TRM/2016.3) (NLM Classification: WB 55.A3).

35. WHO (2016c). Sui Generis Legislative Framework for Protection of Indigenous Knowledge in African Traditional Medicine. WHO Regional Office for Africa, Brazzaville, Republic of the Congo (Document AFRO/EDM/TRM/2016.4). (NLM Classification: WB 55.A3).

36. WHO (2007). Guidelines on good manufacturing practices (GMP) for herbal medicines. WHO, Geneva.

37. WHO (2004). WHO Guidelines on Safety Monitoring of Herbal Medicines in Pharmacovigilance Systems, World Health Organization, Geneva.

38. WHO, 2016d. Regional framework for regulation of traditional medicine practitioners, practices and products (document AFR/EDM/HTI/2016.5).

39. WHO (2004). Tools for institutionalizing traditional medicine in health systems in the African Region. WHO Regional Office for Africa, Brazzaville. Republic of Congo (Document, AFR/TRM/04.3).

8

Role of Partnership in Sector Potential Realization- SDG16- need for Africa-wide Cooperation, Inter-Ministerial Cooperation

Mavis Boakye-Yiadom[1], Zhou Jephias Redemptor[1], Francis Tetteh[1], Irene Aasam Aabeinir[1]

[1]*Clinical Research Department, Centre for Plant Medicine Research, Mampong-Akuapem, Ghana*

Corresponding author: Mavis Boakye-Yiadom

Email: magaby2752001@gmail.com

8.0 Introduction

The UN's Addis Ababa Action Agenda, 2015 stated that "Knowledge-sharing and the promotion of cooperation and partnerships between stakeholders, including between Governments, firms, academia and civil society, in sectors contributing to the achievement of the sustainable development goals should be encouraged". For herbal medicine to contribute to the realisation of SDG-16, there is a need to promote entrepreneurship, and this includes supporting the setting up business incubators and provision of conducive regulatory environments that are open and non-discriminatory.

Private business activity, investment and innovation are major drivers of productivity, inclusive economic growth and job creation. As a result, to promote entrepreneurship in herbal medicine in particular, linkages must be fostered between multinational companies and the herbal medicine industry to facilitate technology development in knowledge and skills transfer, with the support of appropriate policies, on mutually agreed terms. The transfer of such resources from multinational companies will revolutionise the fledgling herbal medicine industry, thus helping in the efforts to unlock its potential. Moreover, businesses within the Traditional and Alternative Medicine (TAM) industry should be encouraged to apply their creativity and innovation to solving health problems and challenges, which will in turn promote sustainable development. In addition, cooperatives or businesses especially, but not necessarily in the industry, should be engaged as partners in the development process, to invest in areas critical to sustainable development, and to shift to more sustainable consumption and production patterns.

Also, it is important to enable the environment at all levels, by ensuring existence of favourable and effective regulatory and governance frameworks. This also entails nurturing science, innovation, dissemination of technologies, particularly to micro-, small- and medium-sized enterprises within the herbal medicine industry.

With the economies of most countries in the African Continental Free Trade Area (AfTCA) especially Ghana, heavily dependent on exporting raw materials, countries can harness the immense potential of herbal medicine, by investing in value addition through processing of natural resources, and productive diversification, while addressing the excessive tax incentives related to these investments.

For Traditional Medicine, Kasilo (2003) noted that some African countries have patented their Traditional Medicine products, while others have been registered by national regulatory authorities (Kasilo, 2003). These new phytomedicines have been added to the global arsenal of medicines for treating some of the most serious health disorders, and contribute significantly to quality healthcare delivery to often poor, rural communities.

The establishment in 1971 of the International Association for the Promotion of Traditional Medicine (PROMETRA), a non-governmental organization, in Senegal is one of the earliest efforts to promote sustainable partnerships in the provision of African Traditional Medicines. PROMETRA conducts scientific research, hosts international conferences and cultural exchanges, publishes a quarterly bilingual journal, Medicine Verte (PROMETRA International, 2016), and coordinates a diaspora-wide network on Traditional Medicine in partnership with academic institutions throughout Africa. These include Noguchi Memorial Institute for Medical Research of the University of Ghana, University of Venda in South Africa, as well as the USA's Morehouse School of Medicine, Atlanta, and North Carolina Central University, BRITE Institute, among others. PROMETRA subsequently established the Centre for Experimental Traditional Medicine in Fatick, Senegal, to conduct traditional medicine research and foster collaboration between traditional health practitioners (THPs) and conventional health practitioners (CHPs), with respect to patient management, treatment and research. This effort was accorded recognition by the Government of Senegal.

In Ghana, the government-owned Centre for Plant Medicine Research (CPMR), which studies and develops herbal products for the management of various diseases, has had a long-standing collaboration with THPs since 1975 (Asiedu-Larbi *et al.*, 2014). The CPMR routinely identifies and consults with THPs to diagnose, prescribe, administer and monitor the effects of selected herbal products on outpatients, and has at least 35 state-registered products for the management of various diseases (Thomford *et al.*, 2014).

8.1 Ubuntu in healthcare

Partnership is indispensable to building strong institutions, as exemplified by the common African concept of *ubuntu*. Although usually used to underline the importance of person-person relationship to the success of an individual and the society at large, the African health industry ecosystem can also fully blossom when there is willingness and full cooperation of each player at every level. In this regard, there might be a temptation to start thinking about partnerships at institutional level while forgetting that the healthcare system is made up of individuals with different backgrounds and qualification. For example, in a system where conventional and traditional medicine are integrated or are being integrated as the case of Ghana, the success of such a system depends more on human interactions. In the case of a healthcare system where herbal and orthodox medicines are integrated, a CHP and THP would need to collaborate in order to achieve optimum patient care. What makes *ubuntu* an apt initiative in this case is the nature of the health care system, in which human beings are key players in the business of working to achieve the unique goal of optimum patient care.

8.2 What are we looking at?

While person-person cooperation or partnership is essential as noted above, the players in the herbal medicine industry have to cooperate with the governing bodies as an institution. Vertical cooperation, where lower levels have to cooperate with higher levels is equally important. For example, in Ghana, Government makes policies through the Ministry of Health, and regulates products in herbal medicine through its agencies such as the Food and Drugs Authority (FDA). However, players such as conventional medicine and its practitioners, are peers (same level) and thus, fall under the term of horizontal cooperation. Apart from that, it is also important to note that the nature of cooperation needed to drive the herbal medicine industry are convoluted, as the players involved have a direct and indirect relationship. While conventional medicine and traditional medicine are easy to link, as they have a direct relationship, by being part of the healthcare system, which falls under ministries of health, some other needed partners such as Agriculture and Finance, have no clear link to health, but can still play a prominent role in influencing the success of the herbal medicine industry. Additionally, the private sector is also known to contribute greatly to healthcare and thus can make a crucial partnership with the traditional medicine industry if need is seen and effort is invested.

8.3 The Traditional Medicine Industry: Where can we start?

The players involved in the traditional medicine industry are many, and this makes it challenging to precisely identify where to start from, and how to establish a worthy partnership. However, it is crucial to try to trace where we are and determine where we want to go. It is also important to determine what the medicine can offer, because partnerships should be beneficial to all stakeholders involved. The herbal medicine industry is already established with existing markets and channels, and will not require building from scratch. What is needed, would be to fine tune the existing channels although new methods of operation.

The practice of traditional medicine is deeply embedded in communities, where everyone may feel entitled to own a part of it. But if everyone owns it, it will be an impediment in the quest to create better value through reformation and moulding. Strategic partnerships are needed in this case to broaden societal impact, and thus improve on the value of the products. This means that the focus should be on fostering consumer confidence, by ensuring that herbal medicines are dependable enough to be considered indispensable to the healthcare system. Every effort must therefore be made to address the question of whether or not we can depend on its quality, availability, safety and affordability. Can traditional medicines garner its deserved plaudits and shift from its complementary role at least to be at par with 'conventional medicines'? If no or yes, respectively, then what kind of alliances do we need to forge in order to achieve the desired goals?

8.4. What are the partnerships needed?

8.4.1 Government policies

Availability of appropriate and relevant policies determines government interest and commitment to the improvement of the traditional medicine industry. When policies and laws are enacted, the needed resources to facilitate research and development of traditional medicine would be easily made available and allocated efficiently. The policies will also guide the practitioners, practice and products to ensure quality and safety.

In Ghana, Government in partnership with the Traditional and Alternative Medicine Directorate (TAMD), has made giant strides in creating an enabling environment for the manufacture, post marketing surveillance, distribution and consumption of traditional medicine products.

8.4.2 Establishment of national traditional medicine research

Although there are several reports on traditional medicine for African countries, abundant opportunities still exist for research, especially in relation to their quality, safety and efficacy. In addition, there is a dearth of human resources with regard to herbal medicine research capability. In collaboration with other relevant Ministries and Agencies, governments should support research into relevant aspects of traditional medicine including, among others, herbs, the role of traditional medicine practitioners in healthcare delivery, technologies used in traditional medicine, the benefits and medical complications/consequences of the remedies, and new formulations of traditional medicines.

Priority should be given to acquiring research capabilities as part of capacity building for implementing policy objectives in this regard. Government must work closely with practitioners in the herbal medicine industry to ensure that there is proper planning, execution and evaluation of research activities so that their outcomes are adopted and applied.

8.4.3 Traditional Medicine Integration

In Ghana, integration by the policy of the Ministry of Health (MOH) is the incorporation of traditional medicine into the mainstream healthcare delivery system at an existing hospital (Hyma and Ramesh, 1994). This involves the introduction of traditional medicines, techniques and knowledge of herbs into the country's mainstream healthcare delivery system. Furthermore, integration denotes the exposure of THPs and CHPs to the philosophies or theories behind TM and orthodox medicine in order to foster cooperation, where an understanding of each other's expertise can allow referrals all for the optimum care of the patient. Bodeker (2001) also noted that an effective integration plan will, among other things, encourage evaluation of traditional medicine in its totality, and ensure equitable resource distribution and arrangement of training programmes for both orthodox and unorthodox practitioners. Thus, integration will seek to achieve a synergy between traditional and conventional medicine, allowing knowledge osmosis with improvement of healthcare delivery as the ultimate aim. Interestingly, whereas THPs have been reported to be willing to partner with CHPs, and even willing to learn and refer their patients to clinics and hospitals, the latter are generally averse to this (Mothibe and Sibanda, 2019). With integration established, one of the first and most benefits to society is that the people will be exposed to much more quality primary healthcare and are less likely to be harmed by the actions of quack practitioners (Good, 1977). The improved quality will spread from the government sector into the private sector since competition will force the private players to improve their standards. All these put together, ensures that the society gains in the end. This will increase the market size for these traditional medications, and significantly reduce funds used to import expensive drugs, which are manufactured wholly or partly overseas. This will significantly reduce the strain on the meager resources of the country.

Integration of traditional health care system will improve the quality of care, giving the patient the freedom to choose and benefit from both health care systems. It will generate much-needed revenue and employment for the people and the nation. For instance, there will be less importation of drugs and recruitment of foreign health personnel. Because medicinal plants abound in large quantities, coupled with the wealth of local knowledge that abound, it will be easier to produce traditional medicines on a large scale and if possible, to export to other countries.

Lack of collaboration between THPs and CHPs, who often disregard the contribution of THPs to healthcare delivery, is one of the main health challenges in Africa. In a study by Kasilo and colleagues in 2019, different countries demonstrated that collaboration between THPs and CHPs can occur in various ways but invariably should result in cohesive teams that are willing to harness each other's potential for the benefit of the patient, the community and the wider society (Kasilo *et al.*, 2019). When both categories of practitioners collaborate, they can enhance practices within the healthcare system, manage collaborative clinics, undertake collaborative research and continue the development of new medicines.

8.4.4 Timely logistics and supply chain management

In Ghana, the supply of medicines though left to private players, is not likely to be monopolized since the list of essential traditional medicines compiled by the Ministry of Health, gives a wide variety of traditional medicines, which are registered by the Food and Drug Authority (FDA). This variety will also ensure that there is a constant supply of medicines since not only one supplier is used. The CPMR in Ghana, undertakes research into traditional medicines for their quality, safety and efficacy.

The current system for procuring plant medicines in Ghana for example, is too bureaucratic. It takes months for orders to be delivered. This may result in untold inconveniences on patient care. If this can be rectified by governments, the traditional medicine sector can see a boost in both its reach and revenues. This can be done by ensuring that the purchase of the herbal medicines is handled in the same way as all other drugs that the facility purchases. An Essential Traditional Medicines list issued by the Ministry of Health will serve as a guide to the procurement department in systems where traditional medicines are part of the mainstream healthcare system.

8.4.5 National Health Insurance Scheme (NHIS) coverage of traditional medicine

Bodeker and Kronenberg (2002) noted that in many countries, the public is financially constrained such that it cannot access traditional medicines because it sometimes has to pay lots of money, as health insurance usually does not cover for these services. Unfortunately, less progress has been made in this regard, as traditional medicine policy has not progressed to include herbal medicines in NHIS. A ministry of health policy that officially recognises frequently used herbal medicines, and includes them in an essential herbal medicines' list would be a great way to start. Medicines covered in the essential herbal medicines' list should be paid for either partially or fully by government under the NHIS so as to make them available and affordable to the average person. If these people could have free access or part payment for some medicines, patronages could increase and the herbal sector will benefit greatly.

8.5 Ministries of Health

8.5.1 Develop and implement health policy

Since the Ministry of Health is responsible for the entire health sector, it is expected to develop health policies in line with Government's overall vision for the citizens. It does this by considering factors that inures to the benefit of the people in terms of quality service delivery. Extending these policies to cover the traditional medicine industry, will yield tremendous results.

8.5.2 Set standards for the delivery of health care in the country

Governments should appropriately regulate the THPs, their practice and products based on standard treatment guidelines. Steps should be taken to ensure compliance with Good Agricultural and Manufacturing Practices, Good Clinical Practice, Good Distribution Practice and any other good practice as may be appropriate. Establishing standards through regulatory mechanisms to ensure the safety, efficacy and quality of traditional medicines and practice, will serve as a restraining order as well as a benchmark to be attained by all health care professionals and the entire population across the country.

8.5.3 National traditional medicine resources development

One of the principal objectives of the National Health Policy in Ghana (Ministry of Health, 2020) is to harness all available human resources inevitably to bring into focus the role of THPs, who shall be identified, screened and appropriately trained in order for them to become effective actors in the health care delivery system. The State Ministries of Health, through the Traditional Medicine Practice Council must establish the structure and process for (Ministry of Health, 2020):

a) Identifying all traditional Medicine Practitioners recognized in their communities.
b) Screening them through registration and accreditation for practicing traditional medicine in accordance with the national guidelines set up by the Traditional Medicine Practice Council.
c) Strengthening their capacity to operate within the framework and national standards for practicing in the health sector.

8.6 Ministry of Finance

8.6.1 Allocation of financial resources

Lack of finance hampers the development of herbal medicines. In a study on researchers at African institutions conducted by Willcox et al (2012), lack of resources (particularly funding, but also lack of infrastructure, equipment, and staff), was the most frequently cited constraint in clinical trials of herbal medicines in Africa. Within the laws governing the practice of traditional medicine, governments, public and private bodies, groups and individuals should be encouraged to establish and finance research or herbal units for the advancement of herbal medicines.

Since the Ministries of Finance are responsible for financial management policy, cooperation with the herbal medicine units will ensure that policies that insulate the traditional medicines industry from funding challenges will be enacted. Adequate funding from government through the Ministry of Finance, should be allocated to ensure proper integration of traditional medicine in national health systems. Funding arising from private sector, technical cooperation, grants, gifts, donations, among others, shall be utilized as appropriate. In Ghana, financial resources shall be made available for public education and information, advocacy, and for training Traditional Medicine Practitioners through both the formal health system and the Ghana Federation of Traditional Medicine Practitioners (GHAFTRAM).

8.7 Private sector

The government is usually the custodian of the health care system of a country as it handles healthcare policies and frameworks. The private sector, can be taken to mean other players outside the government who provide healthcare or healthcare related service. These include but not limited to technological companies, Non-Governmental Organizations (NGOs) and insurance companies. The traditional medicine industry can benefit greatly and accelerate its development if there is a strong partnership with the private sector.

As part of generating employment and decent work for all, in the nation's national development strategies, inclusion of full and productive strategies can also encourage the full and equal participation of women and men, including persons with disabilities, in the formal labour market. Most micro-, small and medium-sized enterprises, which create the vast majority of jobs in many countries, including those in the Traditional and Alternative Medicine (TAM) industry, often lack access to finance. Working with private actors and development banks, will help promote appropriate, affordable and stable access to credit to micro-, small and medium-sized enterprises (SMEs), as well as provide adequate skills development training for all, particularly the youth and entrepreneurs. An example is that of the Agricultural Investment Banks' loans for farmers especially in rural areas. This, if also extended to SMEs or Start-Up projects in the TAM industry, will enable everyone to benefit, and serve as a key instrument for meeting the needs and aspirations of young people.

8.8 Agriculture and Technology

At the core of the Traditional Medicine practice are plants, which are products of farming. Partnership between the traditional medicine industry and the agricultural sector can be a backbone to the former's success. Much of medicinal plant farming, has often been in the hands of small-scale farmers in homes. Agriculture and its technologies can be used to improve the yield and quality of medicinal plants. Studies have shown that the active constituents found in medicinal plants can be affected by the quality of care and environment in whic they are grown (Liu et al., 2016). This means that medicinal plants have to grow as normal crops under the care of appropriate personnel such as agriculturists. This would be essential as better harvest means there is no shortage and there is a broader population outreach. By partnering with the traditional medicine industry, the agricultural technological industry can have a market for their products and thus gain income as well. Additionally, there may be the need for botanists and agronomists who can help in plant innovation. These innovations could include cross breeding of medicinal plants to enhance the level of active constituents, or farming methods that may enhance yield.

With the fast-growing interest in African Traditional Medicine, particularly herbal medicines, where wild harvesting of plant materials is common, involvement of the Agricultural Ministry to encourage the planting of almost extinct medicinal plant species, is essential. This can go a long way to preserve plant species, and also reduce the fast-growing deforestation in the country. Various farmers could be encouraged to establish arboreta, and then monitored on regular basis to ensure compliance with good agricultural practices. This will not only preserve forests and lands, but also provide jobs for people. It will also maintain the existence of important medicinal plant species.

8.9 Scientific Community

The issue with traditional medicines has never been about whether they work; this is evident and that is why they are widely used. The emergence of new infections and limitations of orthodox drugs have created a gap and gullibility among users, resulting in overusage, which can cause toxicity. The world today clamours for evidence-based and efficacious use of medicines. Thus, it lies in the hands of the scientific community to advance the use of the traditional medicines. As has been repeated earlier, knowledge of plant medicine, has largely been passed on through oral tradition, through generations, and much of it is lost on the death of the custodians of that knowledge. Thus, this information can be adulterated and people may use herbs for diseases other than for those they were originally meant to treat. Scientific records will help to preserve the practice. Partnership with institutions of higher learning and of research, can help in advancing the practice of traditional medicine. Perhaps one example is the case of the Madagascar COVID-Organics, or CVO which was strongly condemned and undermined by a section of the scientific community (Daniel Finnan, 2020, Nordling, 2020). Without a rigorous scientific support, it is a challenge for even the most vocal proponents of traditional medicine to make a case for the usage of these products regardless of the testimonies backing them. Herbal medicines are largely associated with nutrition (Walker, 2006) which is evident from their categorization in the US as dietary/botanical supplements as pronounced in the Dietary Supplement Health and Education Act of 1994 (DSHEA). However, some herbs are very potent, such as *Serenoa repens* which was shown to be effective for the treatment of benign prostatic hyperplasia, comparable to the drug finasteride (Wilt et al., 1998). Research institutions created to promote and rationalise traditional medicine usage such as the CPMR in Ghana, Centre for Traditional Medicine and Drug Research in Kenya and The Institute of Traditional Medicine in Tanzania, have a large role to play in pushing the practice of traditional medicine forward. These research centres can also partner amongst themselves, exchanging both research personnel and products. This is particularly crucial as they are located in different African countries, where the flora differs and thus can benefit them both. Tertiary institutions such as the Herbal Medicine Department at Kwame Nkrumah University of Science and Technology and The Africa Centre of Excellence in Phytomedicine Research and Development (ACEPRD) at University of Jos, Nigeria, can also create partnerships which can help innovate traditional medicines. These partnerships can include joint lobbying, roundtable discussions to suggest favourable policies for traditional medicines or exchange programmes which can allow osmosis of knowledge. Evidence-based medicine is very essential to ensure that the patient is safe from harmful use. It is also important to point out that, the scientists in the traditional medicine industry should collaborate among themselves. With the scientific community coming into the fray, there could be innovations in terms of drug formulation, better packaging to improve shell life and drug delivery methods.

The important role of public finance and policies in research and technological development should also be considered. Both public and private venture funds should invest in diverse sets of projects to minimize risks and capture the upside of successful enterprises. This will lead to discoveries of potent medicines for managing most of the alarming health issues.

8.10 Information Communication and Technology (ICT) sector

Information and communication have a prominent role to play in promoting rational use of traditional medicines. The consumer needs to be informed about how to find reliable information and practitioners of traditional medicines by partnering with application makers and technologists, for example, to develop applications that can propagate or disseminate information. An example is the HerbList– an app launched by National Institutes of Health's National Centre for Complementary and Integrative Health. There are also African start-ups on traditional medicines such as Rafamall Inc, which are making in roads, utilising the internet and social media to sensitise people on the scientific use and revolution of traditional medicines. If more partnerships with private ICT players are established, useful sources of information can be made available to the population. It would also be easier to promote new products, improve commerce as products can be dispersed over the internet. ICT can help open the market for newly-manufactured herbal products. Pertaining to traditional medicines, much of the airwaves in many countries in Africa, are dominated by adverts, which tout medicinal plants as virtual 'cure alls'. This has triggered outrage from the scientific community, which views it as a disservice to the strides being made to provide quality healthcare. Additionally, there is the damaging effect of the advent of traditional practitioners without an actual scientific background, but have appropriated scientific elements and symbolism to lure customers dissatisfied with orthodox practitioners (Amoah et al., 2014). Although scientifically inclined practitioners are clearly dominated on the media landscape, they should not be carried away with the wave of excitement and gullibility of the people in their desperation for 'cure all' solutions. Rather, it is important for them to stay grounded in science, for their patients' safety and the integrity of their profession. A proper partnership with the ICT community, will have a role to play by creating a balanced, self-correcting mechanism to mitigate this emerging challenge.

9.0 Conclusion

The herbal medicine sector is green, despite having existed for centuries. For the majority of the world, it has suffered from lack of government support and a skepticism, which has made it seem unfashionable and unworthy of integration into the formal healthcare system. Nevertheless, demand for herbal medicines has remained fairly constant in Africa, with even the industrialised world recently experiencing a resurgence in interest. The persistence of the herbal medicine sector, and its potential benefits, should jolt the various stakeholders into action. As shown in the foregoing discussion, the herbal medicine industry cannot fully develop on its own. However, with government and corporate world support, there is hope. It is also essential to note that, while, we may clamour for partnerships, it is realistically impossible that all partnerships will be achieved. The current discussion seeks to highlight that, the herbal medicine industry is a vast field from which any interested body can play a part and in turn reap some rewards from it. Above all, any partnership should aim at ensuring that the herbal medicine industry contributes to attainment of a better healthcare system- a system which caters for both the poor and the rich, and indeed a system that contributes to prolonging the lives of humanity.

References

1. ACEPRD History. http://aceprd.unijos.edu.ng/history/ (Accessed 3 January 2022).
2. Amoah SKS, Sandjo LP, Bazzo ML, Leite SN, Biavatti MW (2014). Herbalists, traditional healers and pharmacists: a view of the tuberculosis in Ghana. Revista Brasileira de Farmacognosia, *24*(1), 89–95. https://doi.org/10.1590/0102-695X2014241405

3. Asiedu-Larbi J, Adjimani JP, Okine LKN, et al (2014). Efficacy studies on Mist Diodia, a herbal preparation for the management of hypertension in rodents. Medicinal and Aromatic Plant Research Journal; 2:6–7.

4. Bodeker G (2001). Lessons on integration from the developing world's experience. BMJ (Clinical research ed.), 322(7279), 164–167. https://doi.org/10.1136/bmj.322.7279.164

5. Centre for Plant Medicine Research (2022). https://www.cpmr.org.gh/about-us/. (Accessed 5 January 2022)

6. Centre for Traditional Medicine and Drug Research (2022). (CTMDR) https://www.kemri.go.ke/centre-for-traditional-medicine-and-drug-research-ctmdr-nairobi/. (Accessed 2 January 2022).

7. Department Of Herbal Medicine. https://pharmacy.knust.edu.gh/node/145 (Accessed 3 January 2022).

8. Dietary Supplement Health and Education Act of 1994 Public Law 103-417 103rd Congress. https://ods.od.nih.gov/About/DSHEA_Wording.aspx (accessed 25 May 2022).

9. Good CM (1977). Traditional medicine: An agenda for medical geography. Social Science and Medicine (1967), 11(14-16), 705-713. Doi: 10.1016/0037-7856(77)90156-1

10. Finnan D (2020). Profiting from Madagascar's herbal 'cure' for Covid: the story behind artemisia. https://www.rfi.fr/en/africa/20200806-profiting-from-madagascar-s-herbal-cure-for-covid-the-story-behind-artemisia-united-states-france (accessed 3 January 2022).

11. Hyma B, Ramesh A (1994). Traditional Medicine: Its Extent and Potential for Incorporation into Modern National Health Systems. In Health and Development, edited by David R. Philips and Yola Verhasselt. London: Routledge.

12. Kasilo OMJ (2003). Enhancing traditional medicine research and development in the African region. Traditional medicine: Our Culture Our Future. African Health Monitor;1:5–18.

13. Kasilo OMJ, Wambebe C, Nikiema J-B, et al (2019). Towards universal health coverage: advancing the development and use of traditional medicines in Africa. BMJ Global Health;4:e001517. doi:10.1136/bmjgh-2019-001517.

14. Linda Nordling, (2020). Unproven herbal remedy against COVID-19 could fuel drug-resistant malaria, scientists warn. *Science*. DOI: 10.1126/science.abc6665

15. Liu W, Yin D, Li N et al (2016). Influence of Environmental Factors on the Active Substance Production and Antioxidant Activity in Potentilla fruticosa L. and Its Quality Assessment. *Sci Rep* **6**, 28591.https://doi.org/10.1038/srep28591

16. Ministry of Health (2020). Ensuring healthy lives for all (Revised Edition, January 2020). National Health Policy, January, 1–46.

17. Mothibe ME, Sibanda M (2019). African Traditional Medicine: South African Perspective. In (Ed.), Traditional and Complementary Medicine. IntechOpen. https://doi.org/10.5772/intechopen.83790

18. NIH launches HerbList, a mobile app on herbal products. https://www.nih.gov/news-events/news-releases/nih-launches-herblist-mobile-app-herbal-products. (Accessed 2 January 2022).

19. PROMETRA International (2016). Medecine Verte. Science and writing for an African renaissance. Scientific Research and Information Journal.

20. Rafamall. https://rafamall.com/ (accessed 2 January 2022).

21. Thomford KP, Edoh DA, Thomford AK, et al (2014) Effectiveness of the combination of Cryptolepis sanguinolenta and Clausena anisata in uncomplicated malaria. International Journal of Chemistry and Pharmaceutical Sciences, 2:1367–70.

22. Walker AF (2006). Herbal medicine: the science of the art. The Proceedings of the Nutrition Society, 65(2), 145–152. https://doi.org/10.1079/pns2006487.

23. Wilt TJ, Ishani A, Stark G, MacDonald R et al (1998). Saw palmetto extracts for treatment of benign prostatic hyperplasia: a systematic review. Journal of the American Medical Association, 280(18), 1604–1609. https://doi.org/10.1001/jama.280.18.1604

24. Willcox M, Siegfried N, Johnson Q (2012). Capacity for clinical research on herbal medicines in Africa. Journal of Alternative and Complementary Medicine, *18*(6), 622–628. https://doi.org/10.1089/acm.2011.0963

9

Critical Factors for Effective Promotion and Sustainability of African Traditional Medicine: Advocacy, Mass Sensitization, Strategic Stakeholder Engagement, Knowledge Documentation, and Intellectual Property

Bernard Kofi Turkson

Department of Herbal Medicine, Faculty of Pharmacy and Pharmaceutical Sciences, College of Health Sciences, Kwame Nkrumah University of Science and Technology, Kumasi, Ghana

9.1 Introduction

The World Health Organization (WHO) defines traditional medicine (TM) as the sum total of knowledge, skills, and practices based on the theories, beliefs, and experiences indigenous to different cultures that are used to maintain health as well as prevent, diagnose, improve, or treat physical and mental illnesses. TM is therefore viewed as a combination of knowledge and practice used in diagnosing, preventing, and treating diseases. It relies on past experience and observations handed down from generation to generation either verbally, frequently in the form of stories, or spiritually by ancestors or, in modern times, in writing (Mokgobi, 2014).

TM is the oldest form of health care known to mankind. Historically, different societies and cultures developed various useful healing methods to treat diseases (WHO, 2000; Cragg et al., 2001; Abdullahi, 2011). In low- and middle-income countries, about 80% of the population patronise TM for many reasons, including belief, trust, proximity, cost and mode of payment (Sato, 2012). Also, between 65% and 80% of the world's healthcare practice involves the use of TM, commonly referred to as complementary and alternative medicine (CAM).

During the past decades, the developed world has also seen marked use of TM for the treatment of various chronic diseases with success (WHO, 2013; Chintamunnee and Mahomoodally, 2012; WHO, 2008). TM has, therefore, demonstrated immense potential in its contribution to modern medicine (Graz et al., 2010).

TM provides healthcare services based on culture, religious background, knowledge, attitudes, and beliefs that are prevalent in various communities. It includes a range of disciplines involving indigenous herbalism and spirituality, such as divination, traditional midwifery, psychiatry, bone setting, and herbal medicine (Mahomoodally, 2013; Helwig, 2005).

The importance of TM as a source of primary health care was first officially recognised by the World Health Organization (WHO) in the Primary Health Care Declaration of Alma Ata (1978), and has since been globally accepted by many nations around the world (WHO, 1978). It is estimated that since 2018, 170 WHO member states have acknowledged its use (WHO, 2019). According to the WHO, TM is an important and often underestimated health resource with many applications, especially in the prevention and management of lifestyle-related chronic diseases and in meeting the health needs of ageing populations (WHO, 2019).

TM makes use of natural products: animal parts, minerals, and plants. Some forms of traditional medicine practices are traditional Chinese medicine (TCM), Ayurveda, Kampo, traditional Korean medicine (TKM), Unani, and African traditional medicine (ATM) (Fabricant and Farnsworth, 2001). It plays a vital healthcare role in many African communities. Accessibility, availability, affordability, cultural acceptance, and sociological values make them a preferred option for many people over conventional therapy treatment (https://www.afro. who.int/photo-story/traditional-healers-broaden-health).

The development and use of ATM has a very long historical background that goes back to the Stone Age (Abdullahi, 2011). It was the main medical system for millions of people in Africa prior to the arrival of the Europeans, who introduced science-based medicine, which was a noticeable turning point in the history of this tradition and culture.

ATM is a holistic healthcare system and perhaps the most assorted of all indigenous healthcare systems. This is not surprising, given that Africa is considered the cradle of civilisation, with a rich biological and cultural diversity marked by regional differences in healing practices (Aone, 2001).

The use of ATM for the prevention and treatment of diseases is widespread across Africa and has become increasingly popular because of the high cost of allopathic medical health care and the expensive pharmaceutical products, which have become unavailable to majority of people. This may also be due to the perceived safety and efficacy of TM and dissatisfaction with conventional medicines (James et al., 2018). ATM has evolved and remains resilient in spite of the much more standardised Western medicine. Today, TM has become an indispensable part of the continent's national health systems (WHO, 2008). Generally, ATM medicine is seen to be more acceptable to local populations, and can therefore contribute to the attainment of universal health coverage (UHC). The WHO has, therefore, encouraged African member states to promote and integrate traditional medical practices in their health system (WHO, 2008).

Interestingly, despite its high patronage, ATM is still shrouded in secrecy, with very little documentation. However, its future looks bright, if viewed in the context of service provision, increase of healthcare coverage, economic potential, and poverty reduction. Formal recognition will hold much promise for the future (Ezekwesili-Ofili and Okaka, 2019; Ozioma and Chinwe, 2019).

For effective promotion and sustainability of ATM, certain critical factors need to be considered. These include advocacy, mass sensitization, strategic stakeholder engagement, knowledge documentation, and intellectual property rights. The following section outlines these factors.

9.2 Advocacy

Advocacy includes any activity that attempts to educate others and raise awareness about an issue. The main aim of advocacy is to influence decision-makers and change policies (Russell and Levitt-Dayal, 2003). Advocacy activities can include public education and sensitisation to influence public opinion, operational research that profess solutions, constituent action and public mobilisations, agenda setting, lobbying, policy implementation, monitoring and evaluation, and related activities (Christian, 2013). Advocacy is an effective tool for promoting and sustaining systems. It gives people a voice in the decisions that affect them, and helps hold society, governments, and institutions accountable to meet specific needs (FG, 2011).

Policies developed with broad participation help governments and institutions provide better systems. They ensure that systems most affected by policy decisions have a voice throughout the policy development process. Advocacy helps bring disadvantaged groups and systems to the table, and work from the bottom up to include new groups and actors. It also brings together local government, civil society, and other partners to create supportive policy environments (FG, 2011).

Advocacy strategies are developed to provide the tools and skills needed to put plans into practice, placing particular emphasis on the use of data. Advocacy efforts include:

- Helping leaders enhance their role in policymaking
- Helping advocates address issues and delivery of services
- Working with stakeholders to address inequities, and
- Supporting advocacy efforts to disadvantaged groups and systems (FG, 2011).

According to WHO, advocacy involves a combination of individual and social actions designed to gain political commitment, policy support, social acceptance, and systems support for a particular goal or programme (WHO, 1995). Such action may be taken by or on behalf of individuals and groups to create conducive conditions for sustainability and acceptability (Nutbeam, 1998). The goals underpinning advocacy is protecting vulnerable or disadvantaged groups, and empowering people who need a stronger voice by enabling them to express their needs and make their own decisions (Scottish Health Service Advisory Group, 1997).

Advocacy has been recognised as one of the three major strategies and a critical factor for effective promotion and sustainability of African traditional medicine (WHO, 1986). At the International Conference on Primary Health Care held in 1978, the 'Alma-Ata Declaration' was made, with a view to achieving the goal of 'health for all'. In 1998, the WHO incorporated a new global health policy 'Health for All in the 21st Century', and set the goal to achieve health security, health equity, increased healthy life expectancy, and ensure access to essential quality health care for all.

Between 1999 and 2012, through advocacy, the number of WHO member states with national policies on traditional medicine has increased significantly. This includes countries better regulating herbal medicines, coupled with the creation of national research institutes to study traditional medicine (WHO, 2013).

The executive board of the WHO advocated and passed a resolution calling on countries to:

- promote the role of traditional practitioners in the healthcare systems of developing countries and
- allocate more financial support for the development of traditional medical systems. The board also urged the medical profession not to undervalue the traditional medical system.

The WHO recognises that modern medical care is unavailable to the majority of the world's poor, and that to fulfil the primary health needs of all the world's inhabitants, it will be necessary to utilise traditional medical systems. In some countries such as Sri Lanka, India, and China, the traditional health system is legally recognised. The WHO also advocates for utilisation of traditional remedies used by traditional practitioners to effectively treat their patients (Ozorio, 1979).

The goals of the WHO TM Strategy 2014–2023 are to support member states in harnessing the potential contributions of traditional medicine to health, wellness, people-centred healthcare, and UHC while also promoting the safe and effective use of traditional medicine through regulation, research, and integration of TM products, practices, and practitioners into the national health system as appropriate. The current focus of the WHO is to develop norms, standards, and technical documents based on reliable information and data to support member states in providing safe, quality, and effective TM services and their appropriate integration into health systems for achieving universal health coverage and the Sustainable Development Goals (WHO, 2019).

Across the Atlantic, the Ministry of Health in Brazil developed a national policy on integrative and complementary practices (www. bvsms.saude.gov.br), while in the Eastern Mediterranean region, five member states reported having regulations specifically for traditional medicine (Qi Zhang, 2018). Member states in the South-East Asia region have pursued a harmonised approach to the education, practice, research, documentation, and regulation of TM (WHO, 2013). In Japan, it is estimated that 84% of Japanese physicians use Kampō (Japanese TM) in daily practice (Moschik et al., 2012). In Switzerland, certain complementary therapies had been reinstated into the basic health insurance scheme available to all Swiss citizens (Swiss Confederation, 2011). Currently, in Africa, the number of national regulatory frameworks on traditional medicine has increased from one in 2000 to ninety-eight in 2018. Also, 109 countries had launched national laws or regulations on TM, and 124 had implemented regulations on herbal medicines (WHO, 2019).

Advocacy is a critical factor for effective promotion and sustainability of African traditional medicine. Advocacy to include traditional healers in basic healthcare services and promote and sustain the practice is based on the global call for the recognition and legitimisation of African traditional medicine and indigenous healing practices because traditional healers have a central role to play in healthcare delivery in the twenty-first century (Marks, 2006). The biomedical world dismisses the knowledge of traditional healers because this knowledge was not given in recognised academic institutions, but the former has been going through rigorous apprenticeship training that makes them masters of their own (Barimah, 2016). Since 2011, there has been an introduction of traditional medicine in the curriculum of health training institutions and the integration of the same in the healthcare delivery system of some countries, including Ghana (WHO, 2013; Appiah, 2012).

Through advocacy effort and campaigns, traditional medicine has successfully been integrated into mainstream global health care, and those long-standing TM efforts have paid off. The World Health Assembly, the governing body of the World Health Organization, formally approved the latest edition of its influential global compendium, which included a chapter on traditional medicine for the first time (Hunt, 2019).

The Catholic University College, Fiapre, Ghana, in active collaboration with the Ghana Federation of Traditional Medicine Practitioners Associations (GHAFTRAM) and the Centre for the Empowerment of the Vulnerable, advocated a role for traditional medicine in the National Health Insurance Scheme (NHIS) to revive the spirit of 'health for all' to bring health care to 'where people live and work' (Barimah, 2016). This was achieved through advocacy action with policymakers, politicians, and health insurance administrators, with a focus on 'small wins' (Akotia and Barimah, 2007; Payyappallimana, 2010).

Before the introduction of the herbal medicine programme in Ghana, the government introduced a policy that sought to establish, among other things, herbal medicine units in selected teaching and government hospitals to serve as an alternative form of healthcare delivery service in Ghana.

The implementation of the above policy brought with it a major challenge pertaining to the placement of trained medical herbalists within the structure of the Ministry of Health. There was no clear-cut policy on the placement of medical herbalists within the health pyramid in Ghana. Consequently, a vast majority of medical herbalists had to abandon the practice and profession altogether or move into private practice with no clear career progression prospects. This state of affairs adversely affected the practice and recognition of herbal medicine in the country as the profession of medical herbalists became less attractive, to the greater disadvantage of the effort of expanding healthcare delivery and alternative medicine in Ghana.

It is against this background that the Ghana Association of Medical Herbalists (GAMH), through the technical assistance of the Society for Managing Initiatives and Leadership Enhancement (SMILE), Ghana, and the financial support of the Business Sector Advocacy Challenge (BUSAC) Fund, advocated for the complete mainstreaming of medical herbalists in the Ghana health services structure. This advocacy effort was to facilitate the integration process and promote and sustain the traditional medicine sector in Ghana. It also created a forum for stakeholders in the health sector to dialogue and harmonise all policies and programmes on the integration of herbal medicine into the mainstream healthcare delivery system, and draw up a comprehensive road map for this integration process.

In 2011, the advocacy undertaken by the GAMH successfully led to the integration of herbal medicine practice into Ghana's formal healthcare system. This successful advocacy led to the posting of over thirty trained medical herbalists to herbal medicine units in fourteen different public hospitals (www.ghanalinks.org/documents).

The health goal in the United Nations (UN) Sustainable Development Goals (SDGs) is to ensure healthy lives and promote well-being for all at all ages. African traditional medicine has great potential to contribute to the UHC and the SDGs, particularly through strengthening its role in PHC.

Successfully advocating for healthcare and public health policies will ensure users of traditional medicine have access to culturally appropriate healthcare services. This will help promote and sustain African traditional medicine.

9.3 Mass sensitisation

Mass sensitisation is an attempt to make one or others aware of and responsive to certain ideas, events, situations, or phenomena. In this case, sensitisation is the process of making users aware of the existence of traditional medicine globally and its impact on humanity to promote and sustain it. On the other hand, awareness is to know that something exists or to understand a situation or a subject at the current moment based on information/experience (Okon and Ahiauzu, 2008). It can also be understood as knowledge of a situation, truth, conscience, knowledge, achievement, understanding, and perception. According to Akpojotor (2016), mass sensitisation awareness is the knowledge or perception of a situation, fact, consciousness, recognition, realisation, grasp, and acknowledgement concerned with well-informed interest or familiarity in a particular situation or development.

The *Microsoft Encarta Dictionary* (2009) defines sensitisation from three perspectives:

- Having knowledge of something from an observation or being told of
- Noticing or realising that something exists or is happening
- Being knowledgeable and well-informed about what is going on in the world or about the latest development in a sphere of activity

Awareness creation through mass sensitisation is the act of making people knowledgeable about the existence of a situation or phenomenon. It is a veritable information tool used to carry out sensitisation activities to the society.

Mass sensitisation is effective in providing first hand reliable information to communities as the information is cascaded. Oftentimes, governments, groups, organisations, and individuals are faced with an uphill task of creating awareness on health issues (Akarika, 2019). This, to a large extent, is dependent on the dynamics of messages in the communication process conveyed through a mass media sensitisation exercise. The media could be a veritable instrument through which sensitisation, mobilisation, and education are achieved. By means of the information they provide, they broaden our mental and economic outlook (Ukpe and Akarika, 2019).

WHO reports that the utilisation of African traditional medicine has increased tremendously over the past three decades, with not less than 80% of people relying on them for some part of primary health care (WHO, 2004). Although naturally derived remedies have shown promising potential, with the efficacy of a good number of them clearly established, many of them remain untested and their use poorly monitored. The consequence of this is an inadequate knowledge of their mode of action, potential adverse reactions, contraindications, and interactions with existing orthodox pharmaceuticals and functional foods. Since safety continues to be a major issue with the use of traditional medical practices and products, it becomes imperative, therefore, that relevant regulatory authorities put in place appropriate measures to protect public health by ensuring that all herbal medicines are safe and of suitable quality (WHO, 2004).

Furthermore, some of the information on TM available to regulators and practitioners, which tend to affect the effectiveness of the TM regulatory process, may not be accurate because of certain operational factors (Mujinja and Saronga, 2021). Mass sensitisation to create awareness of regulations among practitioners, improve the capacity of regulators, and provide credible information would facilitate compliance with regulations (Mujinja and Saronga, 2021). Mass sensitisation can serve as a means of attracting consumers, researchers, and institutions to take up the development of the industry, with a view to promoting and sustaining it.

The availability of health information and enlightenment campaigns are obligations of the mass media. Lack of awareness has generally led to the negative perceptions some elements in the global community have of African traditional medicine. In the quest to sensitise the global community on the need to promote and sustain traditional medicine, some nations, through the effort of their governments and agencies, outlined several measures to improve the practice. However, despite these measures, there is the need for mass sensitisation to help promote and sustain the practice.

9.4 Strategic stakeholder engagement

Strategic stakeholder engagement is the systematic identification, analysis, planning, and implementation of actions designed to influence stakeholders. It is a continuous and systematic process to establish constructive dialogue and fruitful communication with key stakeholders. Stakeholder engagement and involvement help to ensure a guideline's acceptability and feasibility to end users. This also ensures that equity and other issues are taken into consideration and developed into policy and practice. This, in turn, may lead to improved adherence to any modality and practices under consideration (Schünemann et al., 2014). The purpose of strategic stakeholder engagement is to convey to decision-makers' expectations and stakeholders' interests for their inclusion in decision-making.

African traditional medicine has been used for centuries to improve well-being, and it continues to play a central role in health care. It draws on the continent's rich and unique biodiversity of aromatic and medicinal plants. It is also a promising industry that African countries can do more to export internationally (Moeti, 2020).

During the COVID-19 pandemic, African traditional medicine took the spotlight, starting with the widespread discussion of COVID-Organics as a potential remedy for the virus. The continent's scientists were urged to study this remedy, with a view to scaling up production if it was shown to be effective (Moeti, 2020).

WHO and Africa CDC have supported this process through the development of a master protocol for clinical trials of traditional medicines for COVID-19 and the establishment of a regional expert advisory committee bringing together experts from across the continent to oversee the study of COVID-Organics and other potential remedies (Moeti, 2020).

Various efforts through strategic stakeholder engagement have built on two decades of action to raise the profile of African traditional medicine. Forty countries now have traditional medicines policies, up from eight countries in 2000, and many have integrated traditional medicine in their national health policies and established regulatory frameworks for traditional medicine practitioners. Academic institutions in twenty-four countries now offer traditional medicine courses to pharmacy and medical students. In seventeen countries, referral pathways are established between traditional and conventional health practitioners, and eight countries are strengthening integrated delivery of conventional and traditional medicine services. In Ghana, availability of integrated services has doubled from nineteen facilities offering these services in 2012 to fifty-five in 2020. Mali and South Africa have established partial health insurance coverage for traditional medicine products and services, thus protecting people from financial hardship in line with action towards universal health coverage (Moeti, 2020). Also, in Ghana, through strategic stakeholder engagement, users of traditional medicine can access herbal medicine units for consultation and laboratory services under the National Health Insurance Scheme (ghanalinks.org/documents/20181/0/ BUSAC, 2015). However, users have to pay out of pocket for herbal medications.

There are now more than thirty-four research institutes dedicated to African traditional medicines. In fifteen countries, public funding is allocated on a regular basis to traditional medicine research. Almost ninety domestic marketing authorisations have been issued for herbal medicines, and over forty such medicines are included in national essential medicines lists. Large-scale cultivation of medicinal plants is also increasing, along with local production of herbal medicines (WHO/Afro, 2020).

These achievements show the significant progress that has been made in the regulation and promotion of African traditional medicine. To build on this, more data is needed on the safety, efficacy, and quality of traditional herbal preparations, as well as stronger enforcement of regulatory frameworks and better platforms to share and safeguard traditional medicine knowledge for future generations. Africa's biodiversity, and so too traditional medicine, is also under threat from climate change, and mitigation measures are needed (Moeti, 2020).

Strategic stakeholder engagement is needed to help identify how African countries can best maximise the huge potential of their traditional medicine endowments. It has been observed that of all the great herbal medicine endowments on the African continent, only about 20% has been explored, with still so much work to be done in terms of research and drug development. Also, governments should create enabling environments for researchers and scientists in the African traditional medicine field to influential positions in the fight against diseases which confront the people (Iwelunmor, 2020).

Strategic stakeholder engagement involving governments, academic and research institutions, practitioners, and the private sector is critical to promote and sustain African traditional medicine. Strategic stakeholder engagement should seek to promote the following:

- Utilise traditional health practices effectively in the formal healthcare system.
- Sensitise traditional healers and encourage them to provide home-based care and support community-based health initiatives.
- Establish an effective and efficient control and monitoring mechanism on traditional medicine practice in Africa.
- Promote joint operational research on traditional medicine.

One of the most important decisions taken by the government of Ghana before the integration of herbal medicine into the national health system, was to develop a framework that will guide and direct stakeholder institutions. Unfortunately, this framework was developed with little involvement of all the relevant stakeholders, which hindered the effectiveness and efficiency of the implementation process.

Strategic stakeholder engagement can lead to the sharing of best practices to effectively promote traditional medicine. It can also help in setting research priorities in specific areas for policy making. In addition, strategic stakeholder engagement is required for the identification of sources of evidence, whether historical, traditional or scientific, which support or validate a particular therapy, as well as the determination of the risk/benefit profile, to promote and sustain traditional medicine. This will encourage knowledge generation, translation, and dissemination by establishing a comprehensive and inclusive approach to TM research and development.

9.5 Knowledge documentation

Knowledge documentation is primarily a process in which knowledge is identified, collected, organised, registered, or recorded, as a means to dynamically maintain, manage, use, disseminate, and/ or protect knowledge according to specific goals. The isolated acts of taking a photograph or jotting down a descriptive note need to be part of a comprehensive, thought-through process to be regarded as documentation (WIPO, 2017). Knowledge documentation can be a useful tool as part of an overall strategy for the protection of knowledge.

Knowledge documentation makes it possible to bridge the gap between know-how and knowledge. It identifies, analyses, and makes knowledge 'digestible'. It also facilitates the transfer of knowledge within a giving setting or environment.

Documentation of traditional knowledge (TK) and traditional cultural expressions (TCEs) has attracted increasing attention in recent years from governments and cultural institutions as well as from indigenous peoples and local communities (IPLCs), in parallel with the growing recognition of the cultural and economic value of TK and TCEs (WIPO, 2015).

Documenting traditional knowledge includes recording it, writing it down, taking pictures of it, or filming it – anything that preserves it in an accessible form. It is different from the traditional ways of preserving and passing on knowledge within a community and can promote or damage a community's interests, depending on how the documentation is carried out. Documentation of traditional medical knowledge (TMK) may be useful for the defensive protection of TM by providing information for prior art searches to preclude illegitimate patents (WIPO, 2015).

TMK has social, cultural, and scientific value and is important for many indigenous peoples and local communities. TK is local knowledge that is unique to a culture and society. It is embedded in the community's practices, institutions, relationships, and rituals. It is the total of the knowledge and skills that people in particular geographic areas possess and that enable them to get the most out of their natural environment. Growing commercial and scientific interest in traditional medicine systems has led to calls for traditional medical knowledge to be better recognised, respected, preserved, and protected.

Mwaura (2008) asserted that indigenous knowledge is knowledge of an indigenous community gathered over generations of inhabiting a particular environment. It covers all forms of knowledge – technologies, know-how, skills, practices, and beliefs – as well as cultural knowledge that encompasses intellectual, technological, ecological, and medical knowledge.

ATM is based on the indigenous knowledge of a given people, a given community, and their experiences in the context of the local culture and environment – it is dynamic and changes with time depending on the prevailing situation. African TMK has spread across the world and into many cultures, maintaining a unique and distinctive character (www.daibio.com.vn/en/). For thousands of years, this traditional system of medicine continues to provide effective health care to the vast majority of people of Africa. In more recent years, ATM has been successfully integrated into Western healthcare delivery systems, especially for HIV/AIDS, malaria, TB, and other infectious and chronic diseases. Sandra Anderson of UNAIDS, South Africa, noted that 'traditional health practitioners occupy a critical role in African societies and are making a valuable contribution to AIDS prevention and care' (www.daibio.com.vn/en/).

The utilisation of traditional medicines and associated medicinal plants has been documented by many authors (Mshana et al., 2001; PORSPI, 1992). However, there are still many indigenous cultures and communities around the world that possess a great store of traditional knowledge about traditional medicine for the treatment of various human ailments, which are yet to be documented.

Studies have proven that indigenous knowledge about traditional medicine is continuously being lost through factors such as acculturation and biodiversity depletion (Soelberg et al., 2015). Documentation of this knowledge is very critical for the promotion and sustainability of ATM. Documentation of traditional medicine is important for a plethora of reasons. It ensures that indigenous cultural heritage is preserved for the use of both present and future generations (Mahwasane et al., 2013).

The use of TM is widespread but highly diverse due to floristic and cultural diversity. TM has a huge impact on local economies and biodiversity conservation. The rich history of use of traditional medicines and innovative utilisation of plants as sources of medicines globally, and particularly within Africa, has been passed down from generations to generations largely as oral tradition (Soelberg et al., 2015). TM is usually practiced in secret, and not documented, with the knowledge only shared with a chosen individual (Khumalo et al., 2018). This means that a great deal of valuable knowledge may be lost when the custodians of this knowledge pass on. This will have a negative impact on health systems and the lives of those who depend on it. It is, therefore, important to document it to preserve its socioeconomic and cultural characteristics (Alves and Rosa, 2005), for the benefit of future generations.

Importantly, as part of the effort for rapid development, promotion, and sustainability of TM, research scientists must show interest in the digitisation of this knowledge (Okine, 2019).

9.6 Intellectual property rights protection

Intellectual property (IP) is the set of exclusive rights granted to intellectual creations to the author or the rightful owner of a work of the mind. Intellectual property has two basic branches: industrial property and copyright (WAHO, 2020). Industrial property is a term that encompasses all concepts related to industry and also trade exploitation, in short, the business world. Industrial property includes patents protecting inventions and industrial designs, which are aesthetic creations defining the appearance of industrial products. It also covers product marks, service marks, and layout designs for integrated circuits, trade names and trade brands, trade secrets, geographical indications, and protection against unfair competition. Copyright is the set of exclusive prerogatives available to a creator over his original spirit work. Copyright protects literary and artistic works, regardless of their literary or artistic quality. Original works protected by copyright may be made available to the public by persons or legal entities benefiting then from connected rights or related rights (WAHO, 2020).

Intellectual property rights (IPR) guarantee the protection of products of human creativity and against unfair competition. Protection consists of resorting to the laws, values, and principles of intellectual property that govern patrimonial prerogatives, deposit procedures, and the relevance of inventions. Intellectual property rights make it possible to document and preserve traditional knowledge. However, protecting the products of human creativity should not be at the expense of human rights.

IPR allow any creator, or author, to enjoy the material and moral interests resulting from any scientific, literary, or artistic production emanating from him. The IPR system creates a legal basis and a legal environment that encourage investment. For instance, the patent promotes the marketing of new products and encourages the creation of new techniques for new products. IP information and IPR cannot be taken in isolation from any geographical indication. Information on IP includes all information that has been published in IP documents or that may be derived from the analysis of classification statistics and includes

- technical information taken from the description and drawings of patented inventions or outdated patents;
- legal information on patent claims defining the scope of the patent and its legal status;
- information relevant to companies from the reference data identifying the holder, the filing date, the country of origin, etc.; and
- relevant information on public policies derived from an analysis of deposit trends to be used by policymakers, for example, in the national industrial policy strategy.

Patents are the most important type of IP protection for medicines. To qualify for a patent, an invention must be novel, inventive, and industrially applicable. A patent grants a set of exclusive rights for a limited time, usually twenty years, that allows the inventor to prevent others from making, using, selling, offering for sale, or importing the patented invention without permission (WIPO, 2015).

Patents based on traditional medical knowledge include that of Maca, a traditional Peruvian food and medicine first cultivated by the Incas, and a patent based on Kava, a medicinal plant first domesticated in Vanuatu. In China, patent law protects new traditional medicine-based products, methods of processing, and new uses of traditional medicine, including herbal preparations, extracts from herbal medicines, foods containing herbal medicines, and methods for preparing herbal formulas (WIPO, 2015).

The need to respect the intellectual property rights of traditional societies over their medical knowledge is referred to in WHO's Traditional Medicine Strategy 2014–2023 (WHO, 2013) and is a legal requirement of the Nagoya Protocol on Access to Genetic Resources and the Fair and Equitable Sharing of Benefits, which entered into force on 12 October 2014. However, there are many cases in which medicinal plant products have been developed without respecting the intellectual property rights of the traditional knowledge holders or indigenous resource rights (Willcox et al., 2015).

IPR convey legal ownership over certain intangible assets, such as artistic works, commercial designs, and pharmaceutical technologies. Common types of IP include patents, copyright, trademarks, geographical indications, and trade secrets.

Holders of TMK can nevertheless face significant obstacles in satisfying the conditions required to obtain a patent, especially the requirements of novelty and inventiveness. Because many herbal medicines have been used for generations, disseminated in local communities, and documented in publicly available sources, they may not qualify for patent protection for lack of novelty. Moreover, because herbal medicines typically comprise natural products in their raw form, it can be difficult to claim that their preparation or processing involves an inventive step. However, drugs derived from natural products usually involve some form of alteration or purification, which may be considered a novel and inventive step, making them eligible for patent protection. A trade secret is information not generally known or reasonably discoverable, through which an IP holder can obtain some economic advantage. Once trade secrets become known, they generally cease to enjoy protection. Traditional medical knowledge holders may choose not to disclose their knowledge and keep it secret. In some communities, traditional medical knowledge is known and transmitted only to individual healers and not to the community at large. Other forms of IP may also have a role to play. Trademarks protect distinctive signs, such as words, phrases, symbols and designs that identify the source of a product. This helps consumers identify products with preferred characteristics, such as a specific brand of herbal medicine. Trademark rights are established through either registration or use in commerce. They have been used to market products based on traditional medical knowledge, such as Truong Son Balsam, a traditional balm based on medicinal plants from Viet Nam. However, while trademarks can help distinguish authentic goods, they do not prohibit third parties from using a traditional knowledge without the trademark or under a different mark. Trademarks cannot be used to protect traditional medical knowledge itself. A geographical indication is another sort of IPR that can help to identify the source of goods. Geographical indications identify products as having characteristics associated with their place of origin. However, although geographical indications can be used to distinguish products based on traditional medical knowledge specific to a location, they cannot protect against the same use that is not associated with a place. The way in which geographical indications are protected varies from country to country, and may require registration or use in commerce. Some countries have adopted special sui generis laws and measures, specifically to protect TMK. For example, Thailand's Act on Protection and Promotion of Traditional Thai Medicinal Intelligence protects "formulas" of traditional Thai drugs and "texts on traditional Thai medicine". Only those who have registered their IPR can research, develop and produce drugs using TMK. At the international level, the international legal instrument on the protection of traditional knowledge negotiated by the WIPO IGC would embody a sui generis approach (WIPO, 2015).

Major issues raised by practitioners on IP protection include the number of years an intellectual property could be protected, the geographical indication, and whether plants could be patented. Intellectual property can be protected for up to twenty years subject to the payment of patent fees. Failure to pay patent fees leads to the forfeiture of the patent. At a seminar on IP held at the University of Ghana in 2016 (IASP, 2016), it was recommended that practitioners should document products and processes to aid patenting. Practitioners were also informed of non-disclosure agreements the University of Ghana had put in place to guide engagements with the industry. Members were informed that plants cannot be protected; however, new breeds and products from these plants can be patented if they are new, useful, and not obvious (IASP, 2016).

Issues related to IP can have an impact on products, practices, and even practitioners. IP may support innovation and provide regulations on TM practitioners.

10.0 Conclusion

The foregoing discussion has clearly demonstrated the growing interest in African traditional medicine. Traditional medicine continues to provide health coverage for over 80% of the world's population, especially in developing countries. In addition, medicinal plants continue to be a formidable source of new drugs, with about 90% of the newly discovered pharmaceuticals, derived from them. However, for the ultimate integration of traditional medicine into national health systems, critical factors such as advocacy, mass sensitization, strategic stakeholder engagement, knowledge documentation, and intellectual property rights protection, must feature prominently in countries strategic plans. It is evident that holistic consideration and implementation of these factors, will ensure the availability of skilled human resource, requisite infrastructure, adequate funding and institutional capacity, which are vital ingredients for wider acceptability and sustainability of this most ancient of healthcare systems.

References

1. Abdullahi AA (2011). Trends and Challenges of Traditional Medicine in Africa. African Journal of Traditional, Complementary and Alternative Medicines. 8 (5S): 115–23. http://www.doi:10.4314/ajtcam.v8i5S.5
2. Akarika DC (2019). Awareness and knowledge of prostate cancer information among men in Uyo metropolis, Nigeria. AKSU Journal of Communication Research (AJCR) 79–91.
3. Akotia CS, Barimah KB (2006). Community Psychology in Ghana: Challenges, successes, and prospects. Journal of Psychology in Africa. No. 16. pp. 173–176.
4. Alves RRN, Rosa IL (2005). Why study the use of animal products in traditional medicines? Ethnobiology, 1, 1–5.
5. Aone M (2001). http://www.blackherbals.com/atcNewsletter 913.pdf.
6. Association of Accredited Public Policy Advocates to the European Union (2013). Understanding Advocacy: Context and Use. http://www.aalep.eu/understanding-advocacy-context-and-use. Accessed 24-11-2022
7. Barimah KB (2016). Traditional healers in Ghana: So near to the people, yet so far away from basic health care system. TANG. Humanitas Medicine, 6(2): 9.1–9.6. https://doi.org/10.5667/TANG.2016.000.
8. Bodeker G, Ong C-K, Burford, Grundy C, Shein K. (2005). World Health Organization Global Atlas on Traditional and Complementary Medicine: WHO, Geneva.
9. Business Advocacy News (2015). More Room for Herbal Medicine. BUSAC. https://ghanalinks.org/documents/20181/0/BUSAC+Business+Advocacy+News+Newsletter_March+2015/702 bd623-c797-46d5-acf4-9298f24aecbc?version=1.2
10. Chintamunnee V, Mahomoodally MF (2012). Herbal Medicine Commonly used against Infectious Diseases in the Tropical Island of Mauritius. Journal of Herbal Medicine, 2, 113–125. DOI: 10.1155/2013/617459.

11. Cragg GM, Newman DJ (2001). Medicinal for the millennia: The Historic Record. Annals of the New York Academy of Sciences, 953, 3–25. DOI: 10.1111/j.1749-6632. 2001. tb11356. x.

12. Dhewa C (2008). Is Traditional Medical Practice in Africa still Community Property? – Lessons. From Zimbabwe. https://dlc. dlib.indiana.edu/dlc/bitstream/handle/10535/2172/Dhewa 104101.pdf?sequence=1&isAllowed=y.

13. Fabricant DS, Farnsworth NR (2001). The Value of Plants Used in Traditional Medicine for Drug Discovery. Environmental Health Perspectives, 109(1): 69–75. DOI:10.1289/ehp.01109s169.

14. Futures Group (2011). Health Policy Project. http://www. healthpolicyproject.com/index.cfm?ID=topics-Advocacy.

15. Graz B, Willcox ML, Diakite C, Falquet J, Dackuo F, Sidibe O, Giani S, Diallo D (exican*argemone mexicana* decoction versus artesunate-amodiaquine for the management of malaria in Mali: policy and public-health implications. Transactions of the Royal Society of Tropical Medicine, 104 (33–41). 10.1016/j. trstmh.2009.07.005.

16. Helwig D (2005). Traditional African medicine. Gale Encyclopedia of Alternative Medicine. https://daibio.com.vn/en/ traditional-medicine-1547.daibio.

17. https://ghanalinks.org/documents/20181/0/BUSAC+Business+ Advocacy+News+Newsletter March+2015/702bd623-c797-46d5-acf4-9298f24aecbc?version=1.2

18. https://www.afro.who.int/photo-story/traditional-healers-broaden-health-care ghana#:~:text=Traditional%20medicines%20play%20 a%20vital,many%20people%20over%20conventional%20therapy.

19. Hunt K (2019). Chinese medicine gains WHO acceptance but it has many critics. https://edition.cnn.com/2019/05/24/health/ traditional-chinese-medicine-who- controversy-intl/index.html.

20. Institute of Applied Science and Technology (IASP), (2016). Medicinal Plant Stakeholders Roundtable. Institute of Applied Science and Technology in collaboration with the Technology Development and Transfer Centre of the University of Ghana. https://www.ug.edu.gh/iast/news-information/medicinal-plant stakeholders-roundtable

21. James PB, Wardle J, Steel A, et al. (2018). Traditional, complementary and alternative medicine use in sub-Saharan Africa: a systematic review. British Medical Journal Global Health. doi:10.1136/bmjgh-2018-000895Google Scholar

22. Khumalo NB, Khumalo SV, Nsindane C (2018). The custody, preservation and dissemination of indigenous knowledge within the Ndebele community in Zimbabwe: A case study of Gonye area in Tohwe, Nkayi District. Oral History Journal of South Africa, 6(1), 1–12.

23. Mahomoodally MF (2013). Traditional Medicine in Africa: An Appraisal of Ten Potent African Medicinal Plants Evidence Based Complementary and Alternative Medicine. DOI: 10.1155/2013/617459.

24. Marks L (2006). Global health crisis: Can indigenous healing practices offer a valuable resource? International Journal of Disability, Development and Education, 53, 471-478. https://doi.org/10.1080/10349120601008688

25. Matshidiso Moeti (2020). African Traditional Medicine Day 2020. Message of WHO Regional Director for Africa. https://www.afro.who.int/regional-director/speeches-messages/african-traditional-medicine-day-2020.

26. Mahwasane ST, Middleton L, Boaduo N (2013) An ethnobotanical survey of Indigenous knowledge on medicinal plants used by the traditional healers of the Lwamondo area, Limpopo province, South Africa. South African Journal of Botany, 88, 69–75. DOI: 10.1016/j.sajb.2013.05.004.

27. Mokgobi MG (2014). Understanding traditional African healing. African Journal for Physical, Health Education, Recreation and Dance, 20(2): 24–34.

28. Moschik EC, Mercada C, Yoshino T, Matsuura K, Watanabe K (2012). Usage and attitudes of   physicians in Japan concerning traditional Japanese medicine (kampo medicine): A descriptive evaluation of a representative questionnaire-based survey. Evidence Based Complementary and Alternative Medicine,13.

29. Mshana RN, Abbiw DK, Addae-Mensah I et al. (2001)., Traditional Medicine and Pharmacopoeia; Contribution to the

Revision of Ethnobotanical and Floristic Studies in Ghana, Science and Technology Press, CSIR.

30. Mwaura P (2008) Indigenous knowledge in disaster management in Africa. Available at:https://www.semanticscholar.org/paper/Indigenous-Knowledge-in-Disaster-Management-in/63079f76629e2bda9ac421d1087baf8e626c6e7a

31. Nutbeam D (1998). Health promotion glossary. Health Promotion International, 13, 349–364.

32. Ozioma E-OJ and Okaka ANC (2019). Herbal Medicines in African Traditional Medicine. https://www.intechopen.com/chapters/64851.

33. Okine KL (2019). Knowledge documentation on traditional medicines is vital. https://www.ghanabusinessnews.com/2019/09/19/knowledge-documentation-on-traditional-medicines-is-vital-professor-okine/.

34. Ozorio P (1979). World Health Organization encourages traditional medicine in the third world. Development Directions, 2(4):16.

35. Patrick Iwelunmor (2020). Stakeholders List Ways to Improve African Traditional Medicine Practice. https://pharmanewsonline.com/stakeholders-list-ways-to-improve-african-traditional-medicine-practice/

36. Payyappallimana U (2010). Role of traditional medicine in primary health care: An overview of perspectives and challenges. Yokohama Journal of Social Sciences, 14, 57–77.

37. PORSPI (1992). *Ghana Herbal Pharmacopoeia*, Policy Research and Strategic Planning Institute, Council for Scientific and Industrial Research, CSIR, Ghana.

38. Qi Zhang (2018). Global situation and WHO strategy on traditional medicine. https://www.worldscientific.com/doi/pdf/10.1142/S257590001820001X

39. Sato A (2012). Revealing the popularity of traditional medicine in light of multiple recourses and outcome measurements from a user's perspective in Ghana. Health Policy and Planning, 27, 625–37.

40. Schünemann HJ, Wiercioch W, Etxeandia I, Falavigna M, Santesso N, Mustafa R, et al. (2014). Guidelines 2.0: systematic

development of a comprehensive checklist for a successful guideline enterprise. CMAJ Canadian Medical Association Journal186, 123–42.

41. Scottish Health Service Advisory Group (1997) Advocacy: A Guide to Good Practice. Scottish Office, Edinburgh.

42. Soelberg J, Asase A, Akwetey G, Jägerv AK (2015). Historical versus contemporary medicinal plant uses in Ghana. Journal of Ethnopharmacology, 160, 109–132.

43. Swiss Confederation (2011). Five CAM methods eligible for reimbursement under specific conditions for a provisional period of six years. Available at: www.bag.admin.ch/aktuell/ 00718/01220/ index.html? lang¼de&msg-id¼37173.

44. Ukpe AP, Akarika DC (2019). The broadcast media as agents of moral orientation for youths in Akwa Ibom State. AKSU Journal of Communication Research, 4, 1-14.

45. WAHO (2020). Harmonised Manual on the Protection and Utilisation of Traditional Medical Knowledge. West African Health Organisation. https://www.wahooas.org/web-ooas/sites/ default/files/publications/2317/harmonised-manual-protection-and-utilisation-traditional-medicaal-knowledge.pdf

46. WHO (2013). Traditional Medicine Strategy 2014–2023 Geneva: World Health Organization, 15–56.

47. WHO. (2004). WIIO Guidelines on Safety Monitoring of Herbal Medicines in Pharmacovigilance Systems. Geneva, Switzerland: World Health Organization.

48. World Health Organization (2000). General Guidelines for Methodologies on Research and Evaluation of Traditional medicine.

49. World Health Organization (2008). Traditional Medicine. WHO Fact Sheet No. 134. Geneva: Available at: http://tinyurl. com/5mrd5. (Accessed 3rd March, 2022).

50. World Health Organization (2013). WHO traditional medicine strategy 2014–2023.

51. World Health Organization (2019). WHO Global Report on Traditional and Complementary Medicine. Geneva: World Health Organization.

52. World Intellectual Property Organization (2015). Intellectual Property and Traditional Medical Knowledge. Geneva. https://www.wipo.int/edocs/pubdocs/en/wipo_pub_tk_6.pdf.

53. World Health Organization (1978). Primary health care: Report of the International Conference on Primary Health Care Alma Ata, USSR, 6–12 September 1978. Geneva, Switzerland.

54. World Health Organization (2006): Constitution of the World Health Organization, Chapter 1, Article 1. Forty-fifth edition, Supplement, October 2006 ed. Geneva. WHO. http://www.who.int/governance/eb/who_constitution_en.pdf. Accessed 20 May 2022.

55. WHO/Afro (2020). African Traditional Medicine Day 2020. https://www.afro.who.int/regional-director/speeches-messages/african-traditional-medicine-day-2020.

EPILOGUE

Natural medicine includes herbalism (phytotherapy), which may be defined as a folk or traditional medical practice based on the use of plants and their extracts to treat the sick. Over the years, several studies have consistently shown that herbalism is the most common form of traditional/complementary and alternative medicine. Herbal medicine (phytomedicine) has, therefore, become a very significant component of modern medical care in many parts of the world.

As outlined in various parts of this book, in some parts of the world, majority of the population continues to rely on their own traditional medical systems to meet their primary health care needs. Among non-industrialized societies, the use of herbs to treat disease is almost universal, with a number of traditions currently dominating the practice. These include western herbal medicine, with its Greco-Roman origins, Ayurveda from India, and Traditional Chinese Medicine.

Plants and their metabolites (phytomedicines) have a long history of use in modern "Western" medicine and in certain systems of traditional medicine. In Africa for instance, it is conservatively estimated that about 80% of the population, depends on herbal medicine for their health care needs. It is also known that a significant proportion of the population in countries with advanced health systems, such as the US, seek alternative or traditional forms of health care.

The importance of herbal medicine in the developing world and its growing popularity in industrialized societies has attracted the attention of the pharmaceutical industry and the medical research community, as an extremely vital resource for making new medicines. Herbal medicines are generally used as a direct source of useful therapeutic agents; raw materials for the development of semi-synthetic drugs; prototypes for the design of lead compounds; and taxonomic markers for the discovery of new drugs. It has been variously reported that the global medicine market, based on invoice price levels, is expected to reach about $1.6 trillion in 2025, with about 25% originating directly or indirectly from plants. Examples of such medicines include morphine, codeine, aspirin, digitalis, quinine and atropine. Plants do not only provide pharmaceuticals, but also nutrition and energy.

Advances in plant physiology and pharmacology have shown that plants produce compounds, which may not necessarily be involved in primary photosynthetic and metabolic activities, but have protective properties against predators including animals, insects, and bacteria. For example, the essential oils produced by aromatic plants have antimicrobial properties, some of which have been found to be as effective as conventional antibiotics. Also, plant-derived compounds such as alkaloids, provide some degree of immunity against attacks by herbivores and insects. It is also known that the production or release of these secondary metabolites, tend to increase when a plant is under attack.

Interestingly, these secondary metabolites elicit pharmacological and/ or toxicological properties in humans and other animals, raising the question as to whether this is a coincidence, co-evolution, or convergence.

A notable example is the tobacco plant's production of the insecticide, nicotine, from which the chemically related neonicotinoid (meaning "new nicotine-like insecticides") group of insecticides (e.g. imidacloprid), is obtained. Neonicotinoids are known to be selective for receptors in insects compared to mammalian receptors. They are less toxic in mammals, but are environmentally toxic, especially for the honeybee and wild pollinators. Intriguingly, it is known that morphine receptors exist in animals such as whales and that many plants produce compounds of the morphine class. Morphine has been found to be a component of the defences of antimicrobial plants, but the question remains as to why the cells of a mammal such as a whale, which may hardly ingest the poppy plant, should have receptors for morphine. In fact, eukaryotic cells are known to synthesise authentic morphine, indicating that endogenous morphine has a role to play in animal cell functions. The widespread expression of morphine in plants, vertebrate and invertebrate cells/organs, indicates a high level of evolutionary conservation of morphine and related morphine alkaloids, as essential chemicals for normal growth and development.

Similarly, the cardiotonic steroid, ouabain isolated from human plasma is reported to be identical to plant-derived ouabain. Endogenous ouabain, which is thought to exhibit physiological functions including regulation of vascular tone and sodium homeostasis, has been isolated from bovine adrenal gland and hypothalamus, and found to be identical to plant-derived ouabain. But it is not known what its physiological role in plants is.

The foregoing discussion illustrates the complex relationship between plants and animals extending as far back as our joint evolutionary history. Hominids have coevolved with plants for millions of years, and our strong relationship with, and understanding of them, has enabled us to harness their nutritional, medicinal, and aesthetic benefits. Science can certainly facilitate further exploration of the plant world, to provide the information needed for identification of novel plant-derived compounds with important medical applications. However, as Chapter 3 reminds us, "our immunity is first and foremost innate and natural, and cannot be supplanted by that provided by synthetic laboratory medication".

The history of phytomedicines has, therefore, come a long way from the virtually universal acceptance of the teachings of Galen, Dioscorides, and Paracelsus among others, to an era where they are extracted or synthesised and characterised in cutting edge laboratories across the world. With the support of modern pharmacology and therapeutics, the field of phytomedicines has evolved from an age when natural remedies were organised alphabetically, and not on any observed beneficial or measured effects, into a multibillion-dollar industry.

Nevertheless, as has been previously mentioned, despite the significant contribution phytomedicines make to health care delivery, they are still not officially recognized in many African countries, as policymakers have largely ignored the sector. Consequently, education, training, and research in this area have not been accorded due attention and support. In Chapter 1 of this book, Boakye-Yiadom and co-workers put it vividly thus: "Phytomedicines are still viewed as esoteric, and with suspicion because of the paucity of credible scientific data and have only been featured in discussions when the demand has made them hard to ignore".

For phytomedicine to gain universal acceptability, the quality of research should be improved, and every effort made to address critical issues such as adulteration, and variability in the bioactive constituents of plants, brought on by source and batch differences. Moreover, a very strong case has to be made as to why plant mixtures are to be preferred over the assurance of selectivity provided by well-characterized active constituents as pure drugs.

This book, *A Contextual Exploration of Phytomedicines' Development in Africa*, is the first in a series of books being prepared to share the perspectives of experts in the field, on the immense potential of phytomedicines.

The book outlines such important issues as research and development, good manufacturing practices, regulation and legislation, stakeholder engagement, sustainability and conservation, the role of partnerships in sector potential realization, and critical factors for effective promotion and sustainability of African Traditional Medicine.

As a follow-up to this volume, a second volume is being prepared based on the therapeutic applications of phytomedicines, with special emphasis on the use of phytomedicines for the management of priority and emerging diseases (e.g. hypertension, diabetes, malaria, sickle cell anaemia, tuberculosis, HIV/AIDS, COVID, Ebola), men's health and gynaecological disorders (e.g. fibroids, infertility), among others. It is hoped that the second volume will provide readers with some useful information on the efficacy and safety of phytomedicines, and their important outspoken therapeutic benefits.

These are interesting times, which call for innovation, honest, mutually-beneficial multidisciplinary collaboration, and adequate investment in research and development to produce acceptable, affordable, and readily available phytomedicines. *A Contextual Exploration of Phytomedicines' Development in Africa* seeks to achieve that.

Dr Kofi Busia
Editor-in-Chief
Journal of Herbal Medicine

INDEX

A

Acacia nilotica 26
Acacia spp 26
Acmella oleracea 39
active compounds 135
active ingredients 144
acyclovir 43
adverse reaction 144
advocacy efforts vii, xviii, 27, *28*, 30-1, 56-8, 208, 211-12, 214-16, 233
Africa vii, x, xiii-xv, xvii, 2, 7-9, 12, 14-15, 24-7, 45, 56-7, 60-1, *63*, 107-8, 149-50, 154, 159, 172, 183-4, 186-8, 210, 214, 220, 223, 231-2
pharmaceutical industry in 14
Africa Continental Free Trade Agreement (AfCTA) xvii, 15-16
African Advisory Committee for Research and Development 28
African Development Bank (AFDB) xvii
African Herbal Pharmacopoeia (WAHO) 54
African Medicines Agency (AMA) 184-5
African potato *see Hypoxis rooperii*
African Summit of Heads of State and Government 30
African TM Day 30

African traditional medicine (ATM): vii, 1-2, 24, 26-7, 30, 32-3, 35, 42, 57-8, 67, 140, 157, 164, 167-8, 188, 212, 214, 216-20, 223, 232
African Union Commission for Social Affairs (AUC) 55
Agathosma betulina 16
agrochemicals 92
Algiers Declaration on Research for Health 28
alkaloids, structure of 73
All-Stakeholders International Conference on COVID-19 (ASSIC-19) xiv
Alma-Ata Declaration 7, 209, 212
Alma-Ata Declaration of 1978 *7*, *163*, 209, *212*
alternative medicine (*see also* traditional medicine (TM)) 2, 58, 60-1, 138, 149, 160, 163, 209, 230-1
Amaranthus graecizans L 36
American College of Physicians 69
amino serotonin 73
analgesic aspirin 80
Annals of Internal Medicine 69
antimalarial 34, 37-8, 59, 62
antimalarial preparations 38-9
antiretrovirals (ARVs) 41
antisickling medicines 45
Artemisia absinthium 26

artichoke 89
Association for the Burkina Society of Ethnopharmacology and Ethnobotany 47
Association for the Promotion of Traditional Medicine (PROMETRA) 47, 51, 156, 192
AU Conference of African Ministers of Health 30
Azadirachta indica 40, 62

B

Balanites aegyptiaca 26
batch numbers 123, 133, 137, 144, 146
Benin 53
Bernard, Claude 74
biodiversity 70-1, 74, 102, 104, 220
biodiversity conservation 81, 83, 97, 100, 104, 224
biofertilizers 91
body, human 73
botanic gardens 85, 97, 99-100
Bryonia spp 26
Business Sector Advocacy Challenge (BUSAC) 215, 220, 229

C

Capacity for Clinical Research on Herbal Medicines in Africa 5
Catharanthus roseus L. 25
caution 144
CD4 *41, 43, 50*
Centre for Diseases Prevention and Control (CDC) 55
Centre for Plant Medicine Research (CPMR) ix-xi, 5, 118, 138
Centre for the Empowerment of the Vulnerable 215
centres of excellence (CoE) 57, 154
chloroquine 38-9, 59-60 *see also* Malarial

chronic diseases, better option for treating 10-11
Cinchona ledgemia 40
circumcisions 169
clinical evaluation 35, 40, 44, 63
clinical research ix, xi, 5, 35-6, 40, 43-4
clinical trials xi, 4-5, 35, 38-41, 46, 50, 55, 64, 219
Cochlospermum planchonii 36, 59
collectors 82, 93-6
Commission to the Council and European Parliament 70
community of practice (CoP) 154
complementary and alternative medicine (CAM) (*see also* African traditional medicine (ATM)) 2, 58, 60-1, 138, 160, 163, 186, 209, 229, 231
Conavil 50
Continens (al Hawi fi'Tibb) (herbalists) 26
contract production 123
conventional health practitioners (CHPs) 32, 43, 49, 51, 66, 163, 168, 186
COVID-19 x, xiii-xiv, xvii, 4, 6, 55, 159, 219
COVID-Organics 219
cross-contamination, avoiding risks of 109, 119-20, 125, 132-3
Cryptolepis sanguinolenta 16, 37-8, 46 see also Nibima
cultivators 82, 94-5, 100
Cymbopogon citratus see lemon grass
Cymbopogon giganteus 45
Cyrenaica 26

D

Decade for African Traditional Medicine 30
diabetes 25, 32, 42, 44
Dioscorea 85

Dioscorides, Pedianus 26, 237
diosgenin 85
disease 74
dissipative structures 72
divination 169
Dopravil 50
dosage forms ix, 108, 113, 141

E

East Mediterranean Regional Office
(EMRO) 52
Ebers Papyrus (Ebers) 26
ecological balance 73-4, 89, 92
ecological engineering 71
Economic Community of West Africa
States (ECOWAS) 53, 64
economic crops 84, 87-8, 93
ecosystems 71-2
electro-sensitivity 71
energy 75
equipment sanitation 125
Esoma Herbal Research Institute 46
essential medicines, compounds for 10
ethical behaviour 176
ethics, code of 174-6
ethnobotany 27, 171
Eugenia caryophyllata 46
evidence xv, 5-6, 28-9, 61, 147, 149,
152-3, 221
evolutionary creativity 72
ex situ conservation 13, 97, 99
excipients 112-13
Experimental Centre for Traditional
Medicine (CEMETRA) 51

F

FACA 45, 62
Fagara 45-6
Fansidar 39-40
farmers 60, 81-4, 86-7, 93, 100, 104 *see
also* cultivators
fertilizers 83, 86, 91

finished herbal products *see* herbal
medicines
'first in, first out' (FIFO), principle
of 119
first plant-derived drug *see* morphine
first semi-synthetic drug pure *see* salicin
Food and Drugs Authority (FDA) 5,
111, 139
food plants 46
funding, public 220

G

general health practices 169
Geneva Foundation for Medical
Education and Research 48
Ghana Association of Medical
Herbalists (GAMH) 215-16
Ghana Federation of Traditional
Medicine Practitioners
Associations (GHAFRTAM) 215
Ghana Herbal Pharmacopoeia 46, 53,
59, 232
Ghanaian herbal medicine industry 5
Global Ministerial Forum on Research
for Health 57
Global Strategy and Plan of Action on
Public Health, Innovation and
Intellectual Property 57
Goldman Sachs Investment Bank 70
good agricultural practices (GAP) 87
good laboratory practices (GLPs) 171
good manufacturing practices (GMP)
13, 108-37, 139-40, 189
green manure 91
Guidelines on Good Agriculture and
Collection Practices (GACP) for
Medicinal Plants 132
Guinex-HTA 45
Gundamiti 42
Gynandropsis gynandra 45

H

healing, longest history of *see* African traditional medicine (ATM)
health care, oldest form of *see* traditional medicine (TM)
health policies, national 7
health systems 8-9, 29, 63, 65-7, 149, 159, 162, 164, 186, 188-9, 211
herbal industries 95-6
herbal materials 112, 135
mixing and blending of 134
herbal medicine industry, revenue generation of 11-12
herbal medicines:
attributes of 9
regulation of 67, 151, 165, 169-70, 182, 184
herbal pharmacopoeia 32, 46, 52-3, 59-61, 64, 232
herbal preparations 38, 41, 43, 52, 108, 112
herbal products 80
herbalists 108
herbs 112
herpes zoster 43
heuristic approach 76
HIV/AIDS: 28, 32, 40-3, 50, 52, 59, 61, 67, 140, 186
fungal infections associated with 43
holistic health 76
holistic model 148
homeopathy 74
humans 71-5, 79-80
Hyoscyamus muticus 26
hypertension 25, 32, 44, 60
Hypoxis rooperi 42-3

I

ibn Basil, Istifan 26
ibn Ishaq, Hunayn 26
improved cultivars, lack of 88
Incas 226

indigenous knowledge x, 57, 67, 163, 189, 223, 231-2
indigenous peoples and local communities (IPLCs) 222
Industrial Revolution 75
Institut Malagache des Recherches Appliquées 44
Institute of Health Sciences Research 44
Institute of Traditional Medicine of the Muhimbili University of Allied Health Sciences (MUHAS) 32, 48-9
'Integrating Modern and Traditional Medicine: Facts and Figures' (SciDevNet) 151
integration, lack of 85
Intellectual property (IP) 224-5, 227-8
inter alia 28-31, 170
International Conference on Primary Healthcare in 1978 *see* Alma-Ata Declaration of 1978
International Plant Protection Convention and Codex Alimentarius 92
International Regulatory Cooperation for Herbal Medicines (IRCH) 150, 161

J

Jatropha gossipiifolia 45
Jatropha gossypiifolia 45

K

Kava 226
Kenya 49-50
Kenya Medical Research Institute (KEMRI) 34, 39
knowledge documentation vii, 208, 211, 221-2, 228, 232
Kwame Nkrumah University of Science and Technology ix, xi, 47

National Health Insurance Scheme (NHIS) 215, 220
National Institute for Medicine Research (NIMR) 50
National Institute for Pharmaceutical R&D 34
National Institute for Pharmaceutical Research and Development (NIPRD) v, ix, 39, 46, 50
National Institute for Public Health Research 32
National Institute of Medical Research 48-9
National Institute of Phytotherapy and TM *see* National Institute for Public Health Research in Mali
national medicines regulatory authorities (NMRAs) 33, 166, 170, 184
national policies 6, 67, 155, 165, 174, 184, 189, 212-13
national regulatory authorities (NRAs) 172, 175, 192
National Research Institute 32
natural products ix-xi, 25, 46, 59, 102, 105, 151, 210, 226
Natural products for Eastern and central Africa (NAPRECA) 29
nature 2, 69-71, 73-5, 78, 103, 106, 138
nature reserves 97-8, 100
Nelson Mandela Hospital 43
Nibima 37-8
Nigeria v, ix, 25, 39, 46, 48, 53, 55, 61-3, 67, 154-6, 229
Nigerian Herbal Pharmacopoeia 53
NIPRD 92/001/1-1 *39*
NIPRD-AM-1 *39*
Niprisan 46
noncommunicable diseases 35, 44-5
nongovernmental organisations (NGOs) xix, 31, 43, 48-50, 97, 156
Novartis 151, 156
nutraceuticals 3 *see also* phytomedicines

O

Onopordon spp 26
organic farming 92
Organisation of African Unity (OAU) xx, 15, 29-30, 57, 62, 185
Ottawa Charter of 1986 *148*

P

P. falciparum 40
packaging materials 136
Papaver somniferum 80
Paracelsus (Swiss physician) 9, 70
parasitaemia 38-40
parasite clearance 39-40
Pasteur, Louis 74
Pausinystalia yohimbe 25
personal hygiene 113, 116-17
personnel training 124
pharmacognosy ix, 124, 171
pharmacopoeia, African 46, 54, 62
photosynthesis 72
phytobiology 72
phytoextraction 74
phytomedicines vii, ix, xiii, xv-xvi, 1-3, 5-12, 14-16, 25-6, 30, 33, 35-6, 42, 44-5, 59, 64, 69-72, 74-6, 108-10, 132, 149
active constituents of 2, 4-5, 89
controlling herbal (plant) materials of 113
controlling non-plant-based starting materials of 116
controlling starting materials of 113
industry of 15
labelling 136, 144, 146
manufacturing of 109, 111, 113
producers of 115, 122, 131
production of 108-9, 132-3, 137
qualification and validation of small-scale facilities of 121
phytomedicines: vii, ix, xiii, xv-xvi, 1-3, 5-12, 14-16, 25-6, 30, 33, 35-6,

www.ingramcontent.com/pod-product-compliance
Lightning Source LLC
Chambersburg PA
CBHW021353210526
45463CB00001B/86